Underground Frequency Guide

A Directory of Unusual, Illegal, and Covert Radio Communications

Third Edition

Donald W. Schimmel

Solana Beach, CA

Printed in the United States of America

Cover design: Brian McMurdo, Ventana Studio, Valley Center, CA
Developmental editing: Jack Ryan, Langley, VA
Interior design and production services: Greg Calvert, Artifax, San Diego, CA

ISBN: 1–878707–17–5
Library of Congress catalog number: 94-078677

Contents

APPENDIXES

Preface to the Third Edition

When Harry Helms and I exchanged letters regarding the possibility of an update of his **UNDERGROUND FREQUENCY GUIDE**, Harry indicated he would really like to see the project undertaken but said that, unfortunately, he did not have time to do it. I wonder if the reason he did not have the time had anything to do with the fact that HighText Publications had just relocated its offices to a building overlooking a wide, sandy beach where some episodes of the television series *Baywatch* were being filmed!

However, I am delighted that Harry asked me to prepare this edition and it is my sincere hope that it will be received with the same acceptance as the preceding editions. In this new version, a few categories have been added which were not in earlier editions. One in particular is a glossary (Appendix C), which I hope will prove helpful in understanding the many special terms used to describe the various elements pertaining to covert radio monitoring, signals intelligence, and espionage. Several categories which were formerly included have been dropped because of lack of coverage or because those signals are no longer heard.

This book does not include any classified information. It is based on the monitoring efforts of shortwave listeners plus analysis procedures contained in various books, including recently declassified U.S. government books listed in Appendix A.

Acknowledgments

This third edition of the **UNDERGROUND FREQUENCY GUIDE** incorporates the loggings of many contributors along with those by the author.

A huge thanks to the following shortwave monitors:

Robert Margolis	Simon Mason
David Sabo	Ary Boender
Gary Hamlin	Tom Sevart
Corey Soderlund	Ken Sooy Sr.
Richard Baker	Brad Low
Harry Helms	George Zeller
Tom Mazanec	Roger Caldicott

There were some other individuals who also contributed material but who wanted to remain anonymous. While it often is not possible to corroborate items from anonymous sources, when such information appeared to be plausible, I included it because it may very well be legitimate and authentic.

Special recognition must be given to Jacques d'Avignon. Through his use of a computer based propagation program and interpretation of the resulting data, Jacques was able to provide valuable leads to possible locations of a few network stations. His efforts were very much appreciated.

Thanks also go to Kevin Tubbs who provided some excellent spectrum analyzer charts.

Dedication

This third edition is dedicated to the memory of Dave (Poco) White. Although Dave and I never actually personally met, we had exchanged letters for many years dating back to when I wrote the "Utility Intrigue" column in ***Monitoring Times***. During this entire period, Dave sent me a large volume of loggings which were used in the various columns I wrote. He always furnished interesting comments pertaining to the intercepted communications.

Dave took great pride in his monitoring and direction-finding efforts, and was always willing to check out something for me. He frequently made tape recordings of signals I was interested in but could not hear well enough to copy at my location.

Dave is now at the big monitoring station in the sky. He passed away in April, 1993. He is sorely missed by his fellow "mystery station" monitors.

Introduction

As has been the case for many SWLs, my shortwave pursuits changed over the years. While my initial efforts were directed to listening to international broadcasting stations, I soon found that I had a yen for following unidentified transmissions. At times, I have covered a particular activity during a period of several years, compiling as much information as possible in an effort to identify the station or network.

Readers should bear in mind that frequencies and schedules do change periodically. Thus, it is very likely that some of the loggings listed herein will no longer be heard on the reported frequencies. By carefully tuning through the spectrum, you may be rewarded with the detection of a new schedule, time, or frequency in use for the signals discussed in this book.

Many of the "underground" activities cannot be positively identified because they operate in a very secure manner with little more noted than the frequency, time of operation, callsigns (if any), and the type of traffic. Also, the absence of both plain language messages and operator chatter eliminates two potential sources of leads. This leaves us with just a few tantalizing details, and consequently we are unable to make a solid determination concerning sponsorship of the monitored communications.

I have indicated my opinions regarding certain intercepted targets. These opinions were formed based on my interpretation of data gleaned from studying collected traffic. I certainly recognize that others may interpret such data differently, but I have tried to be as thorough as possible in arriving at my conclusions.

This edition of the UNDERGROUND FREQUENCY GUIDE contains a listing in Chapter Five of stations and signals observed from 1992 to 1994 by listeners in North America as well as by some listeners located elsewhere in the world. I believe the loggings represent a balanced sample of "underground" transmissions.

There are a number of titles listed in Appendix A. These provide further technical details and/or revealing background information concerning the activities described in this book. I urge readers to check out some of the references given as I am sure they will prove to be of interest.

1

Numbers Stations

FOR MANY SHORTWAVE RADIO FANS, the so-called "number stations" *are* underground radio. While there are many other unusual things that can be heard on shortwave, there's no doubt that numbers stations are the most widely heard type of underground radio and the first unusual signals you're likely to notice. In fact, it's almost impossible not to run across numbers stations if you do any tuning whatsoever outside the conventional shortwave broadcast or amateur radio bands.

Some Background

"Numbers stations" are so named because that's what you'll hear: groups of statistically random numbers, usually in blocks of four or five digits, read by a machine-generated voice similar to that used by the telephone company to announce changed or discontinued numbers. A female voice is usually used to announce these numbers (although a male voice is rarely used), and the languages you're most likely to hear are Spanish, English, and German. However, exceptions abound to the "typical" numbers transmission just described. Sometimes the broadcasts go out "live" instead of being machine-generated. On rare occasions, both male and female announcers have been used or two different languages, such as Spanish and German, have been mixed in the same transmission.

Probably due to the abundance of espionage books in recent years, it now seems that most numbers buffs do believe numbers stations are used to transmit enciphered messages to espionage agents operating under cover. Also, it appears to have been established by such books that many numbers messages are most likely enciphered by the *one-time pad* (OTP) method.

Substituting a book (held by both parties) in place of cipher pads also provides a source of literal keying text which while somewhat similar to the OTP principle is nevertheless a much less secure form. For detailed explanations of several variations of the OTP system, it is recommended that readers refer to ***The Code Breakers*** by David Kahn. In addition, a number of titles which describe reported use of OTP systems, are mentioned in following pages.

The cipher groups in some numbers messages, according to various espionage books, may be subjected to cryptographic arithmetic (See "additive" in the Glossary) to deter opposition efforts at cryptanalyasis. A couple of book ciphers are briefly described on pages 121-123 of ***Solving Cipher Problems*** by Frank W. Lewis. For those who may be interested in titles pertaining to cryptographic and intelligence subjects (in addition to those listed in the References section), it is suggested you request a catalog from Aegean Park Press, P.O. Box 2837, Laguna Hills, CA 92654. Included in their catalog are a number of declassified U.S. Government COMINT training texts which have comprehensive treatments of many cryptographic systems.

Another system that may be in use with some numbers transmissions is the so-called "dictionary code." This is possibly the system in use with the "three/two" numbers messages where there is a clear and distinct pause between the third and fourth digits of each five-digit block. A book (or an actual dictionary) is used which is available to both the sender and recipient of the traffic. The first three digits would represent the page number of the book with the second two digits representing the position of a word on the page (usually, but not always, counting from the upper left corner of the page).

A dictionary type system, whether it is actually a dictionary or a book, must have a feature called a "speller" to take care of words which would not be found in the volume being used as the code book. For example using Webster's New World Dictionary, 2nd College Edition, a proper name like SMITZER would appear in the coded version as 1250/1 846/20 693/6 1446/4 1651/15 437/9 1168/16. Sometimes dictionary codes utilize an additive (see the Glossary section) to disguise the true location of the plain text words.

One example of a "spy" numbers broadcast was the traffic sent from Moscow to London for KGB agent Gordon Lonsdale. He was assigned a three-character call sign which included a figure 1 if the broadcast was live traffic. If the 1 was not present, it signified it was dummy traffic.

- In *The Spy Who Got Away* by David Wise (Random House, 1988), the training of CIA defector Edward Lee Howard is detailed. In chapter seven, "Moscow Rules," there is a discussion of how the CIA contacts its agents. It is stated that the CIA transmits messages to its agents using five-digit number groups which are decoded using the one-time pad method. It is also stated the CIA broadcasts many numbers messages intended for no one; the CIA just wants the KGB to think it has many more agents operational than is actually true!

- *Widows* by William R. Corson, Susan B. Trento, and Joseph J. Trento (Crown, 1989) tells of two espionage agents who received instructions via shortwave radio. One was Nick Shadrin, a Soviet defector who eventually went to work for the CIA and became, at the CIA's urging, a double agent for the KGB. In chapter 13, "Shadrin: Tightrope," the authors tell how Shadrin received messages via shortwave radio. The messages were groups of numbers in the Morse code transmitted from Cuba by an operator known as "Michelangelo" because of his skill in sending the code. The second was Ralph Sigler, a career Army sergeant who was sent by Army intelligence as a double agent to infiltrate the KGB. Sigler made his approach at the Soviet embassy in Mexico City, and was quickly accepted by the KGB. Chapter 17, "Sigler: The Operation" describes how Sigler also received messages in Morse code over shortwave radio. There is also a description of how Sigler decoded the messages.

- ***KGB Today: The Hidden Hand*** by John Baron (Reader's Digest Press, 1983) tells in chapter seven, "Devoted Agent," how the KGB provided one agent with "...a little brown cube oscillator to be fitted into a Hammerlund (sic) HQ180 radio to improve shortwave reception in Canada..." (This sounds more like the crystal-control unit that was available for the HQ180; it allowed reception on several fixed frequencies for which users had installed crystals.) The messages received from the KGB supposedly took a long time to decipher (up to 18 hours in one case) and were often inconsequential, such as May Day greetings or reminders that it was the anniversary of the establishment of the Germany Democratic Republic. In chapter nine, "The Man Who Loved To Spy," Baron tells the story of Professor Hugh Hambleton, who spied for the KGB while a teacher at Quebec's Laval University. He received instructions from the KGB over a Grundig receiver equipped with a device known as a "luminere." This was a device for agents unwilling or unable to master Morse code, and converted the Morse characters into individual numbers from 0 to 9 using what seems, from the book's description, to be a crude nixie tube or electric lamp readout. All messages were transmitted in five-digit blocks, which Hambleton deciphered using a one-time pad.

Other books, such as the paperback edition of ***The Puzzle Palace*** by James Bamford, also carry details of agents receiving messages as coded number groups. All these examples don't mean that each and every numbers message is necessarily from the CIA, KGB, or DGI (Cuban Intelligence Service), but no alternative explanation is backed up by anywhere near the amount of evidence as the espionage hypothesis.

In 1988 I wrote a two-part article, "Secrets of Shortwave Espionage," which was published in the May and June issues of ***Popular Communications*** magazine. The material was later summarized and appeared in the "DX Reporter," the monthly bulletin of the Association of DX Reporters, a SWL club. I have since revised the summary somewhat and believe it appropriate to include material from it throughout the remainder of this chapter.

Espionage Insights

Suppose you were going to set up a clandestine system of communications with your agents in the field. It seems logical that the overall traffic would be composed of three main categories:

First, there would certainly be some practice traffic. This would be required to accustom the trainees to listening for and tuning in their respective broadcasts so as to receive traffic addressed to them. This practice traffic would provide valuable experience and improve their proficiency in copying messages as well as in decryption or decoding, as the case may be.

A second type of traffic could very well be deception traffic. What better way to confuse opposition traffic analysis and cryptanalaysis efforts than by throwing in traffic that was nonsense! Dummy messages could be made up with normal headings, but with texts of purely random digits, as just one of the ways of generating non-valid traffic. The absence of a pre-specified indicator group would tip off the agent that the traffic was bogus, and he need not finish copying it. Eventually, the overall plan should probably include a phasing out of the practice messages after a suitable breaking-in period of the agent and replace those messages with dummy traffic and then blending in the live traffic.

The third type of traffic would, of course, be valid messages. I doubt that every message would be of the "BLOW UP THE DAM" class. Instead, there would probably be the myriad of details to be accomplished such as selection of secure meeting places, instructions regarding financial matters, notification of mailing addresses, crypto details, directions concerning meeting schedules and means of recognition/identification of the agent handlers. Any of the thousand and one real life concerns of the spy organization and its undercover personnel overseas. This getting settled-in may require many months, and in some cases perhaps years, before all of the necessary preliminary assignments have been carried out. I would suspect that only then would the agent be activated to provide reporting on a regular basis. Bear in mind that it stands to reason that all the time he is carrying on his clandestine activities, it would seem imperative that he also maintain a normal profile with some type of work plus engaging in other appropriate daily activities that "fit in," thus assisting in building and maintaining his cover.

A possible reason for so many one-way voice broadcasts is that perhaps it's just simply not a good idea from a security standpoint for the agents to possess transmitting equipment. The solution, therefore, is to provide a common type radio, one with SW bands, which would certainly not arouse any suspicion when viewed by agent acquaintances. This may also explain why some numbers broadcasts are in MCW, since a communications type receiver with a BFO is not required.

When not permitted to use radio for agent to headquarters traffic, a possible reply channel could be via mail where an innocent appearing text serves as a means of conveying secret writing or a message in an open code system.

For security purposes, such mail would probably be sent via an indirect route. Perhaps first to a P.O. box address where it would be picked up, outer wrapping replaced, and then remailed to the final destination address. This lengthy time en route could explain why certain messages are repeated for weeks or months at a time. It could be simply because the agent handler at the spy organization headquarters has not received an acknowledgment from the agent in the field which indicates receipt of a particular message.

Lastly, let us consider the cryptographic system to be employed. It goes without saying that not only do we want a good secure system, but one which is also not too complex to use. Doesn't it make sense that any spy outfit worth its salt is not going to use some rinky-dinky system?

PHONETICS FOR VARIOUS LANGUAGES

	English	*Spanish*	*Portuguese*
A	Alpha	Alfa	Antena
B	Bravo	Brasil	Bateria
C	Charlie	Canada	Condensador
D	Delta	Delta	Detector
E	Echo	Espana	Estatico
F	Foxtrot	Francia	Filamento
G	Golf	Guatemala	Grade
H	Hotel	Hotel	Hotel
I	Item	Italia	Intensidade
J	Juliet	Japon	Juliet
K	Kilo	Kilo	Kilo
L	Lima	Lima	Lampada
M	Mike	Mejico	Manipulador
N	November	Noviembre	Negativo
O	Oscar	Oscar	Onda
P	Papa	Papa	Placa
Q	Quebec	Quito	Quadro
R	Romeo	Radio	Radio
S	Sierra	Santiago	Sintonia
T	Tango	Tango	Terra
	English	*Spanish*	*Portuguese*
U	Uniform	Universidad	Unidade
V	Victor	Victor	Valvula
W	Whisky	Whisky	Watt
X	X-ray	Xilofono	Xilofon
Y	Yankee	Yucatan	Yucatan
Z	Zulu	Zulu	Zulu

	French	*German*	*Italian*
A	Alfa	Anton	Alfa
B	Bravo	Berta	Bravo
C	Charlie	Casar	Canada
D	Delta	Dora	Delta
E	Echo	Emil	Europa
F	Foxtrot	Friedrich	Firenze
G	Golf	Gustav	Guatemala
H	Hotel	Heinrich	Hotel
I	India	Ida	Italia
J	Juliet	Julius	Juventus
K	Kilo	Konrad	Kilometro
L	Lima	Ludwig	Lima
M	Mike	Martha	Messico
N	November	Nordpol	Novembre
O	Oscar	Otto	Otranto
P	Papa	Paula	Palermo
Q	Quebec	Quelle	Quebec
R	Romeo	Richard	Romeo
S	Sierra	Siegfried	Santiago
T	Tango	Theodor	Tango
U	Uniform	Ulrich	Universita
V	Victor	Viktor	Venezia
W	Whiskey	Wilhelm	Whisky
X	X-ray	Xanthippe	Xilofono
Y	Yankee	Ypsilon	Yokohama
Z	Zulu	Zeppelin	Zelanda

NUMBERS AS SPOKEN IN VARIOUS LANGUAGES

	English	*Czech*	*German*	*Yiddish*	*French*
0	zerio	nula	null	nul	zero
1	one	jeden	eins	ein	un
2	two	dve	zwei	tsvei	deux
3	three	tri	drei	drei	trois
4	four	ctyri	vier	fier	quatre

	English	*Czech*	*German*	*Yiddish*	*French*
5	five	pet	funf	finef	cinq
6	six	sest	sechs	seks	six
7	seven	sedm	sieben	sibben	sept
8	eight	osm	acht	acht	huit
9	nine	jevet	neun	nein	neuf

	Italian	*Serbo-Croatian*	*Spanish*	*Arabic*	*Portuguese*
0	zero	nula	cero	sifr	zero
1	uno	jedan	uno	wahid	um
2	due	dva	dos	itsnayn	dois
3	tre	tri	tres	tsalatsa	tres
4	quattro	cetiri	cuatro	arbaa	quatro
5	cinque	pet	cinco	chams	cinco
6	sei	sest	seis	sitta	seis
7	sette	sedam	siete	sabaa	sete
8	otto	osam	ocho	tsamaniya	oito
9	nove	devet	nueve	tisaa	nove

(The following entries are phonetic.)

	Russian	*Mandarin*	*Cantonese*
0	nol	LING?	ling!
1	adin	YEE	YAT
2	dva, dvie	AHRR!	yee
3	tri	SAHN	SAHM
4	cetire	SER!	SEIGH
5	pjat	woo?	ng?
6	sest	LEEOH!	look
7	siem	TSYEE	CHAT
8	vosiem	BPAH	BAHT
9	djevit	jyoh?	GOW?

Chinese pronunciation notes: Capital letters indicate high pitch
? indicates rising pitch
! indicates falling pitch

There are other tone variations, but inclusion of them is beyond the scope of this brief listing.

Cut Numbers

Some numbers broadcasts are in CW or MCW Morse code. A method of reducing on-air time for Morse number messages in through the use of "cut numbers." The Morse code for certain letters are substituted for the longer, numbers Morse code. There are many of these cut number systems in use throughout the world. Here are the breakouts for a few that have been observed.

This is a familiar one where the only number cut is the zero:

1 2 3 4 5 6 7 8 9 0
1 2 3 4 5 6 7 8 9 T

A minor variation of that one involves the use of the letter O in place of T for the zero.

This one is based on the typewriter keyboard:

1 2 3 4 5 6 7 8 9 0
Q W E R T Y U I O P

The 4F group numbers broadcasts use this one:

1 2 3 4 5 6 7 8 9 0
AUV4 E 6 BDNT

An expansion of that system looks like this:

1 2 3 4 5 6 7 8 9 0
AU3 4 5 6 7 DNT

And we can't overlook the one seen in some of the 5F numbers traffic:

1 2 3 4 5 6 7 8 9 0
ANDUWR I GMT

Lastly, here is one seen in maritime communications:

1 2 3 4 5 6 7 8 9 0
AUV4 5 6 BDN T

Broadcast Formats

Let's now take a look at some different types of numbers broadcasts. Several have been included which, while rare catches for North American monitors, are heard by others located elsewhere. It is also very interesting to note identical types of broadcasts being heard in different areas of the world. This certainly must indicate the involvement of a large, wide-spread organization.

Not all numbers transmissions take place smoothly. There have been boo-boos observed related to number message transmissions. I recall noting several myself, including one involving three different broadcasts on the same frequency and at the same time. Another time a compromise occurred when a five-digit numbers broadcast ended but the carrier remained on the air. A few minutes later the introduction for the Radio Havana shortwave programming commenced. It was only on for an instant and then abruptly pulled. Then there was a real puzzler for a few moments until I realized that I had caught a tape being played backwards.

Five-Digit Spanish Numbers by a Female Announcer

This is the most commonly heard type of numbers station in North America. Each transmission opens with a female voice repeating the Spanish word "atención" followed by a three-digit number block and a two-digit block, as in "atención 843 68." The three-digit block is believed to be either the recipient of the message or some information concerning the one-time paid (such as page number or position to start the decoding process). The two-digit block is always the number of five-digit groups making up the message. This opening sequence is repeated for a few minutes prior to the actual message. The number blocks end with the words "final, final." In some cases, the message may be repeated. Most of the transmissions are in AM, although a handful of SSB transmissions have been reported. The technical quality of many five-digit Spanish transmissions is low. You'll hear lots of hum and noise on them, even when the signals are strong, and the audio is frequently "mushy" and distorted.

A variation of five-digit Spanish numbers stations follows the same pattern noted above, but adds an additional two-digit and three-digit block following the opening sequence. Using the previous example, "atención 843 68" would be repeated for several minutes at the opening. When this stopped, a sequence such as "49 180" would be repeated for about a minute before the start of the actual five-digit groups. In this case, the last three digits ("180") would be the number of five-digit groups. This format is much less common that the first one. There are some very rare formats, such as one that follows "atención" with one three-digit group and two two-digit groups. Since these are so seldom heard, I won't discuss them here; Havana Moon's book has a full treatment of them.

The groups making up some Spanish five-digit messages may be repeated consecutively, as in "04107 04107 08027 08027 37715 37715 70760 70760," etc. (This also happens on some English five-digit transmissions.) It is not known what significance, if any, this might have; it could just be insurance to get the message across to the intended recipient if reception conditions are expected to be difficult.

One interesting quirk involving the Spanish five-digit stations with a female announcer is how a message that beings on or near the hour may be repeated on the half-hour on a frequency a few kHz away from the frequency used for the broadcast on the hour. It's not known what the purpose of this rebroadcast is; one obvious theory is that it is simply "insurance" in case the intended recipient missed or had difficult copying the original message. However, there is no direct evidence supporting this or other explanations.

A quick scan of the frequency listings in the last half of this book will show you how common five-digit Spanish stations are. If you listen during your local evening hours, it's likely that you'll hear several of them in a single evening. Don't be too surprised if you find an open carrier (i.e., dead air) on the frequencies used by these stations for a long period of time before a numbers transmission begins. Open carriers on some frequencies have been noted for more than an hour before a numbers message finally starts!

Four-Digit Spanish Numbers by a Female Announcer

This is the second most commonly heard type of numbers station in North America. These typically open with a three-digit number group repeated three times followed by a count from one to zero, as in "834 834 834 1234567890." This is repeated for several minutes and is usually terminated by a sequence of ten tones, followed by the word "grupo" and a two-digit group representing the number of four-digit groups in the message. This is sent twice, and then the four-digit number groups begin. At the end of the number groups, the word "fin" is sent. Often, the message is repeated. While these transmissions sound like AM (and are so listed in this book), they are often actually USB with enough remaining carrier to allow them to be copied on AM receivers. This particular mode is referred to as *reduced carrier sideband,* abbreviated as RCS or sometimes as RCSSB. The technical quality of four-digit Spanish transmissions (and, indeed, all four-digit messages regardless of language) is uniformly excellent.

Five-Digit Spanish Numbers by a Male Announcer

Far rarer than the previous two types of Spanish numbers stations, these transmissions are generally identical to those of the five-digit type using a female announcer. One tendency, however, is that many of these stations apparently transmit "live" messages (although they may be taped in advance). Unlike the precise machine-generated number blocks used by the female stations, there have been receptions of these stations where the male announcer has hesitated, made an error and corrected himself, and even coughed!

Five-Digit English Numbers by a Female Announcer

These were once very similar to the Spanish five-digit transmissions, opening with "attention" followed by what was believed to be the addressee and the group count; each message was terminated by "end." More recently, these simply begin with a three-digit block repeated for several minutes and then directly into five-digit groups with no indication of the group count. Some people have reported that these transmissions are preceded by the letter "N" in Morse code prior to the start of the actual message, although I have never heard this.

In recent years, some five-digit English numbers stations have started using a tune at sign on known as the "Lincolnshire Poacher," which to American ears sounds like the old "Pop Goes The Weasel" tune. This particular station is widely heard in Europe and eastern North America.

One curious aspect of the five-digit English stations (and the three/two variety discussed later) is how the accents of the female announcers vary. The female announcer of signals on 7519 kHz had a distinct British accent, for example. Others faintly "roll" the letter "r" in "four" and "zero" as if their native language were Spanish. Most have a neutral accent such as that found in the American mid-west or west coast. (These variations may also be present in other languages; if your native tongue is Spanish or German, or if you speak either well enough to detect accents or variations, you might want to be alert for such differences.)

Four-Digit English Numbers by a Female Announcer

The call up is a three-digit number block repeated for several minutes, followed by a 1 - 0 count and then four-digit groups. Sometimes beeps or tones precede the message texts. The technical quality of these stations is high. Warble jamming directed against these broadcasts has been noted.

"Three/Two" Digit Spanish Numbers

These are very similar to the five-digit Spanish numbers stations, but there is a clear and deliberate pause between the third and fourth digits of each number block. The pause is approximately equal to the time required to announce one number. Most of these messages use a female announcer.

Five-Digit German Numbers by a Female Announcer

These stations are distinctive because they often open with some sort of musical signal, such as march music, synthesized tones, a flute, or music box. Some stations open with a two-letter phrase from the international phonetic alphabet, such as "Papa November." Five-digit German stations use "achtung" instead of "atención" and "ende" in place of "final." Following "achtung" is a three-digit block believed to be the addressee of the message. This is followed by "gruppen" and the count of the number blocks in the message. In most other respects, these are similar to Spanish stations although the technical standards are consistently higher.

Four-Digit German Numbers by a Female Announcer

These are comparatively rare and seldom heard in North America. They are similar to the four-digit Spanish stations, but often use the same type of sound effects and music at the opening as the five-digit German stations.

"Three/Two" Digit German Numbers

These opened with a three-digit number group followed by a 1 to 0 count in German. This was followed by ten beeps, "gruppen" then the message group total, and finally the message itself. The Papa November station of this broadcast network went off the air in October 1992. Related stations DFC37 and DFD21 ceased operation in December 1992. As of early 1994, some of these stations were still heard.

"Three/Two" Digit English Numbers

These open in a manner similar to the four-digit English stations, although a series of beeps or tones and the word "count" followed by the number of groups in the message usually precede the actual number groups. The message is normally repeated twice during each transmission. The technical quality of these stations is likewise high. An increasing number of these transmissions has been heard in North American since late 1988.

Morse Code Numbers

Like "verbal" numbers stations, Morse code stations transmit messages in groups of four or five numbers. And, as with other numbers stations, the four-digit variety tend to have higher engineering and technical standards than the five-digit variety tend to have higher engineering and technical standards than the five-digit ones. Most of these stations transmit "pure" CW, meaning the unmodulated transmitter carrier is turned on and off to form the Morse code characters. To receive these stations, you'll need to tune with your receiver's mode selector set to "CW" or turn your receiver's beat-frequency oscillator (BFO) on. A few broadcasts use modulated CW (MCW), in which the Morse characters are sent in the AM mode by using audio tones to form the characters. Many CW numbers stations use some form of "cut numbers" in which letters of the alphabet are substituted for numbers. This is done because letters are shorter Morse characters than numbers.

Miscellaneous Numbers Stations

As a quick examination of the frequency listings in this book will show, numbers stations are sometimes heard in such languages as Russian, Bulgarian, Rumanian, Yiddish, etc. West coast and Pacific listeners sometimes hear numbers stations in Chinese. There were also reports in the early 1980s of "mixed language" transmissions involving both German and English or German and Spanish number groups. In addition to the variety of languages there are many, many different formats for the numbers broadcasts.

Null Messages

These CW broadcasts often consist of just a three-digit identifier repeated three or four times followed by five consecutive zeroes, as in "875 875 875 00000." Thus leading to the belief that the five zeroes signify there is no traffic upcoming. Five zeroes have also been seen at the end of a message and in this instance they probably stand for "no more traffic." Reportedly, transmissions with three and five zeroes are controlled by the KGB and its successor organizations.

Bulgarian Betty

This broadcast led off with an interval signal of a five note rising scale tune, followed by the group headings. Messages were sent in the five figure format. It disappeared for a period and then was again reported in operation as of August 1993 with a daily schedule commencing at 1355 on 5311 kHz.

OLX Station

Here is another overt call sign being used, reminiscent of the now defunct DFC/DFD call signs. In this case, OLX is assigned to CETEKA Press, Prague and it would certainly seem to be a foregone conclusion that the numbers traffic has no relation to a press function. The call up is "VVV DE OLX" in CW.

The broadcasts have been monitored at 0000, 0200, 0300, 0400, 0500, 0600, 0900, 1100, 1300, 100, 1900, 2000 and 2100. Frequencies included 5301, 6758, 8142 and 14977 kHz. The messages have been 5F groups in CW and by a YL/Czech also.

The Ten Minute Numbers Broadcasts

These MCW broadcasts featured four daily schedules with very short messages, 4-5 groups, of four-digit blocks sent for exactly ten minutes. The schedules were 0030 on 5264/6792 kHz; 0230 on 6840/9958 kHz; 1030 on 7725/10324 kHz; and at 1830 on 11491/16310 kHz. A study of a quantity of these short messages seemed to indicate the distinct possibility this was a deception program with dummy traffic transmission.

There was also a YL/SS voice broadcast which displayed a similar format with some groups repeated from day to day. The schedule was 0230 daily and in the AM mode. This broadcast also ran for exactly ten minutes.

DEAR FRIEND !

RADIO STATION **OLX** THANKS YOU FOR YOUR REPORT ABOUT

RECEPTION OF ITS TRANSMITION ON THE 26. may 1994

AT 13.00 GMT ON FREQUENCY (IES) 8142

kHz

OUR ADDRESS IS :

Dear Simon!

TKS REPORT AND PHOTO

VY 73!

MINISTERSTVO VNITRA CR
P. B. 21/SK
170 34 PRAHA 7

OUR RADIO STATION **OLX** IS TRANSMITING ON FREQUENCIES

o - FREQUENCIES USED DURING SUMMER TIME
x - FREQUENCIES USED DURING WINTER TIME
PARTICULARS ARE IN kHz.

3239	3280 O + X	3333	4601 X	4757	5301 O + X
6280 X	6758 O + X	6865	6958	7577	8142 O + X
9353	10125	10307	11002 O + X	11416	11585
14977 O	15897	16046	18303	20865	22910

Figure 1-1: Simon Mason of England managed to get this QSL for a numbers transmission from station OLX in Prague, Czech Republic.

Three-Letter Phonetic Alphabet Stations

These stations can be thought of as the "first cousins" of numbers stations. They use a female voice to repeat a phrase composed of letters from the international phonetic alphabet, such as "Kilo Bravo Alpha Two." These transmissions are repeated for hours at a time, but are sometimes interrupted for messages consisting of groups of letters from the phonetic alphabet. These transmissions started to be widely heard in the late 1970s.

An article in the July 1984 ***Popular Communications*** by Greg Mitchell claimed that these transmissions are produced by the Mossad, the Israel intelligence service. Mitchell presented compelling evidence (based upon a visit to Israel) that some of these stations do indeed transmit from Israel, and the Mossad theory is currently the most widely accepted explanation as to their origin. However, it's clear that not all of these transmission originate within Israeli territory. Some of these stations are heard in North America at times and on frequencies which would be unlikely to support propagation from Israel. Some monitors in the metro New York area have reported powerful, local-like reception of these signals that indicated the transmitter was nearby. (In one case, a listener in Hoboken, NJ, was able to hear such a station "pinning" his receiver's signal strength meter although he had the antenna on his portable shortwave radio fully collapsed!) Israeli embassies and consulates around the world could house some of these transmitters. Another possibility, given how the Mossad will sometimes cooperate closely with other intelligence services, is relays from friendly nations.

Jeff Heyman of New York, NY, devoted considerable time in 1989 and 1990 to monitoring these stations. He mainly listened to 10125 kHz, where he had heard a station identifying as "Charlie India Oscar X-ray Two." Jeff heard it throughout the evening in New York, and noted it usually began at 45 minutes past the hour and ran for about ten minutes. (He also noted other unusual things on 10125 kHz between such transmissions, including "beeping," RTTY signals, and, once, bizarre "gypsy" music followed by a man yelling in an unidentified language.) Beginning in the late summer of 1989,

Jeff started hearing variations. During the last two weeks of August, "Charlie India Mike Two" was heard instead on 10125 kHz. It changed back to "Charlie India Oscar X-ray Two" in early November. When it returned, the phrase was repeated only from 45 to 50 minutes past the hour.

Things got stranger in 1990. On January 22, Jeff heard "Charlie India Oscar Two" repeated *continuously* on 10125 kHz. On January 23, he heard what sounded like "Charlie India Oscar Tango (Rome?) (Bill?) Bravo (Zero or Zebra) (Rome?) Bravo (Rome?) Bravo (Five?)." This was repeated continuously all evening. On January 24, "Charlie India Oscar Two" was again heard continuously throughout the evening. But on January 25, "Charlie India Oscar Two" returned to its previous schedule of 45 to 50 minutes after the hour.

February, 1990 saw numerous bizarre changes in the operating pattern on 10125 kHz. On February 4, Jeff heard "Charlie India Oscar Uniform Eight," and he also heard the same signal on 6745 kHz. On February 7, the message on 10125 kHz seemed to change on an hourly basis. The most unusual one heard was "Charlie India Oscar Tango One Sierra Mike One Four Bravo Six Zero Zero," which ran for over two hours. Other messages heard that night included "Charlie India Oscar Two" and "Charlie India Oscar Three." On February 8, "Charlie India Oscar Three," "Charlie India Oscar Oscar," and "Charlie India Oscar One Delta" were among the calls heard. The next night, the only signal heard was "Charlie India Oscar Two" at 45 to 50 minutes past the hour, and this pattern continued throughout the rest of 1990.

So Where are Numbers Stations Located?

In the past, it was generally assumed that four-digit messages originated within the United States or its allies (such as West Germany and Britain) while the five-digit stations were believed to originate from Cuba (for Spanish and English stations) or East Germany (for German and English stations). There was some powerful evidence to back this up, including the fact that Radio Habana Cuba was sometimes heard mixing with five-digit Spanish numbers. This implied that both originated at the same transmitter site, and that "crosstalk" or intermodulation between the two signals was happening there. In 1978, a Florida listener named David Crawford contacted the FCC about some five-digit Spanish signals he was hearing on 3060 and 3090 kHz, and was told by the FCC that they were coming from Cuba. Finally, the notion that the technically crude, sloppy five-digit messages originated from the "other side" while the clean, technically precise four-digit signals were "ours" had a certain appeal to many American SWLs.

The first definite "fix" on a numbers station location within the United States came in 1984. A mathematics professor at a small Connecticut college determined that four-digit Spanish numbers groups were coming from a Warrenton Training Center station located near Remington, VA. I received a phone call from the individual shortly after his discovery, and as I recall the conversation, he indicated he had used a portable shortwave receiver and "homing" techniques. This involved observing signal strength as he drove around, and by getting closer and closer, he finally was able to pinpoint the location of the target transmissions. A sign at the station showed it was affiliated with the National Communications System (NCS).

At the time it was thought this station was part of the Vint Hills Farms complex. However, I later received information from an anonymous individual who had been at the Vint Hills base, and he said while it used to be an Army Security Agency intercept station in past years, it had not functioned in that capacity for quite some time.

He further stated the only major activity remaining at Vint Hills was a research and development function and that was due to be relocated because of the anticipated complete closure of the Vint Hills base. Lastly, he claimed that while the Warrenton stations were part of the NCS structure, Vint Hills Farms station was not.

With the 1984 discovery, some people considered the "numbers mystery" solved. The FCC said it had determined that five-digit signals came from Cuba. Four-digit signals had been definitely tracked down to northern Virginia. The popular theory had been confirmed, and the case was closed, they thought.

However, not all numbers stations receptions could be fitted into this theory. In particular, listeners in southern Florida were able to hear four-digit Spanish numbers on 4670 kHz during the daytime at a strong level —a most unlikely reception if the signals were coming from Warrenton. Other listeners noted signals that were clearly "single-hop" receptions from only a few hundreds of miles away, and had to be coming from transmitters closer than Virginia or Cuba. The receptions in south Florida prompted SWLs in that area to search for a transmitter site. The June, 1988 issue of ***Monitoring Times*** magazine carried a report by columnist Larry Van Horn which presented strong evidence for a four-digit transmitter site near Jupiter, FL. Later, another potential transmitter site near the Miami Zoo was located.

Some of the most tantalizing yet controversial evidence on the locations of various numbers stations came in late 1989 from an individual known pseudonymously as "Havana Moon." Havana Moon is the author of a book on numbers stations, and has also written about them for such magazines as ***Popular Communications* and *Monitoring Times***. On October 16, 1989, Havana Moon noted a five-digit Spanish station on 6577 kHz at 0205 UTC. This quickly became an active channel for five-digit transmissions and attracted the attention of many SWLs since 6577 kHz is also used for aeronautical communications by aircraft in eastern North American and the Caribbean. SWLs were able to hear pilots complain of interference from the signals. Since there were legitimate public safety considerations involved, Havana Moon says he reported the signal to the FCC monitoring station at Vero Beach, FL, and asked them to locate its source. Havana Moon claimed the FCC took direction-finding "fixes" on November 5 and 20, 1989, and that the signals were coming from a site near Esteli, Nicaragua. The FCC supposedly took another direction-finding fix on December 11 and found the 6577 kHz signal was now originating from near Jamelteca,

Honduras. Havana Moon asked the Vero Beach monitoring personnel to make additional measurements, and they reportedly found that five-digit Spanish and CW numbers transmissions on 3927 kHz were coming from Guineo, Cuba and that five-digit Spanish transmissions on 3690 kHz were originating from Havana itself.

Unfortunately, there has been no independent verification of Havana Moon's claims. Havana Moon himself remains anonymous, and the FCC personnel at the Vero Beach facility have refused to publicly confirm Havana Moon's claims. Havana Moon's reputation was badly damaged in 1993 by the collapse of a SWL book and equipment business known as "The Radio Collection" he was associated with. Havana Moon and his partners in the venture abruptly disappeared when it closed, leaving substantial sums of money owed to both suppliers and customers. While Havana Moon's 1989 claims seemed like a major breakthrough in the numbers stations mystery at the time, Havana Moon's subsequent conduct has called his credibility into severe question. And even if Havana Moon did accurately report what he had been told by FCC personnel, there is no way to assess whether the FCC personnel were giving him accurate information or just deliberately trying to confuse the issue. What seemed to have been a promising development in late 1989 has now turned into a frustrating dead end.

Monitors in Ohio and California have observed strong local-like signals from numbers stations. Many more signals are of the "single-hop" variety from within a few hundred miles. Almost any installation used by the U.S. military or civilian agencies for high-frequency communications could potentially be used for numbers transmissions. George Zeller, a well-known DX expert from Cleveland, began paying close attention to a five-digit English station on 5046 kHz (usually parallel to 6840 kHz) during June, 1990. During June, local sunset in Cleveland is approximately 0100, yet George was able to hear the 5046 kHz signals daily from 2200. At that time, the signals were good (about S8) and also the *only* signal that George could hear on 60 meters at the time. George asked other listeners, via the Association of North American Radio Clubs (ANARC) amateur radio net, to listen and report their receptions. One who did was Joe Buch, N2JB, then in Virginia Beach, VA. N2JB reported hearing 5046 kHz at 1522—that was 11:22 a.m., his local time—with very strong signals. Since 6840 kHz had been identified as one of the frequencies used by the Warrenton facility, George felt this was a strong indication that the 5046 kHz signals also originated at Warrenton. While that is certainly a reasonable conclusion, the Norfolk area is loaded with government transmitting facilities (especially those belonging to the U.S. Navy) and a facility closer to N2JB cannot be ruled out.

George Zeller speculated further about those transmissions in a letter to Harry Helms. "Let's assume 5046 is in Warrenton—evidence is piling up for this assumption," wrote George. "Now, why would they want to transmit numbers on 5046 in broad daylight from Warrenton with high-powered transmitters that put a good signal into Cleveland? This certainly is a worthless frequency/time choice for any reliable reception over 300-500 miles. The target area must also be in the eastern United States (or possibly southern Ontario or Quebec, which I doubt). Why would anybody want to transmit numbers from Warrenton to this target area? The whole concept baffles me. It seems like a totally worthless expenditure of taxpayer money. We have telephones, fax machines, two-way radio of various types, and loads of secure methods of communication in the USA. Why use numbers? I can think of no reason at all. The main effect of a 60 meter daytime numbers transmission from Warrenton is that DXers like you and me and N2JB can figure out the QTH of the numbers transmitter. It seems to have no other purpose. Or are we perhaps missing something about the real purpose of USA numbers transmitters in general?" In reply, Helms speculated that one possible answer was that such transmissions were for training purposes, and the use of on-the-air signals was to help trainees learn how to properly use a shortwave receiver.

Another possibility is that some numbers transmissions could be coming from foreign embassies and consulates. Under international law, embassies and consulates have the right to maintain radio stations for contact with their home nations. In reality, there are no restrictions whatsoever on what these stations can be used for, and some could easily be used for numbers messages. While embassies are in Washington, consulates are in major cities scattered about the country. While most consulates are little more than a suite of offices, some (such as the Russian consulate in San Francisco) are every bit as elaborate and well-equipped as any embassy. In his book ***Breaking with Moscow***, Soviet defector Arkady Shevchenko tells of radio communications facilities in use at the San Francisco consulate as well as at the Soviets' United Nations mission in New York, their U.N. mission residence in New York, and at their retreat at Glen Cove, Long Island. In addition to the Soviets, many Warsaw Pact nations were known to have radio communications facilities within the United States. (The fate of these stations, in view of the collapse of the Soviet block, is uncertain.) It is also widely suspected that the Israelis maintain radio communications capabilities within the United States.

Average SWLs can play a big part in solving this puzzle. The advent of a new generation of high-performance portable shortwave receivers means it is easier than ever to track down a transmitter site—if the receiver overloads on a signal when near a transmitting facility, then the signal is coming from that transmitter site. (This is how the transmitters in Virginia and Florida were found.) Patient SWLs with portable receivers could likely unearth some fascinating things if they live near major metropolitan areas or military installations.

A new generation of direction-finding loop antennas is also coming into wide use among DXers. While direction-finding with loops at shortwave frequencies is always somewhat inexact (sophisticated "Adcock" antenna systems are necessary for precise "DFing"), these new loops do permit a significant "narrowing down" of possible transmitter locations.

Anomalies

Given their general air of mystery, it is not surprising that quite a few "anomalies" have been observed in connection with numbers stations.

Harry Helms observed a couple on February 9, 1990. He was listening to 6785 kHz at 0501, and heard a five-digit Spanish station begin its transmission with "atención 687 02" repeated by a woman in the AM mode. Seconds after the transmission began, a high-pitched "whine" similar to that produced by a digital counter circuit began. This sound quickly rose in volume and intensity, changing into a sound similar to a badly corrupted ARQ phasing signal. The signal was about one S-unit stronger than the numbers station. At 0508, the interfering signal abruptly stopped but there an open carrier remained, causing a heterodyne to the numbers station. The interference resumed at 0511 and continued until the numbers broadcast ended at 0519, at which point the interference also stopped. While this was taking place on 6785 kHz, Harry had a second receiver tuned to 7845 kHz. At 0502, he heard a female voice in the AM mode start repeating "atención 877 03" and then five-digit groups at 0505. At 0509, some distorted Spanish pop music was heard in the background of the numbers station. This continued until 0516, when a series of data bursts began. These were strong enough to obliterate the numbers station when they were sent. At 0518, some scrambled speech in the USB mode was also heard atop the numbers station. This continued until 0520, when both the numbers station and interfering signals left the air. There have been several other cases of apparent jamming of numbers stations, usually with a so-called "warble" jammer, and several of these cases are described in the Underground Frequency List of this book.

Another anomalous reception by Harry Helms took place on September 8, 1990. At 0505, he was listening to a "three/two" digit Spanish station on 6824 kHz. As he monitored the signal, Harry became aware that the audio from English language telephone conversations were mixing with the numbers station. The calls were brief and nondescript, initiated by the same man, and had a strong similarity to phone patches handled between armed forces members and their stateside families on military affiliate radio service (MARS) frequencies. The calls were of a personal, inconsequential nature, in a "I'm fine, I hope you're doing well" vein. There were dialing tones and rings between calls, but no operator intervention or assistance. At one point, the man initiating the calls told the party he reached to "be careful what you say; we're on the radiotelephone and someone might be listening." Calls lasted less than two minutes each, and ended when the numbers station transmission ended at 0514. The carrier remained on the air, however, and Helms continued to monitor. At 0523, the main returned with a call to a woman which lasted less than a minute. Nothing further was heard until 0530, when the carrier left the air. Helms was able to hear this signal on four different receivers, so the chance of it being the result of a receiver mixing product or image is highly remote.

Monitors such as Brian Webb of California and Zel Eaton of Missouri have noted five-digit Spanish numbers stations above 30 MHz during periods of sporadic-E propagation. Propagation on frequencies above 30 MHz is typically limited to "line of sight" (the optical horizon plus another 20% or so due to bending of signals around the horizon). However, during certain periods intense patches of ionization capable of propagating VHF signals over distances up to 1500 miles can develop in the E layer of the ionosphere; this phenomenon is known as sporadic-E. The numbers stations operating above 30 MHz are easy to spot because they are in the AM mode while almost all other "normal" signals about 30 MHz will be in the FM mode. Checks of possible fundamental frequencies have yielded nothing during such receptions, so it appears these transmissions may be studio to transmitter links of some sort. Numbers stations have been reported above 30 MHz during periods of sporadic-E propagation that produced receptions from the southeastern United States through northern South America.

One promising avenue of investigation involves the use of spectrum analyzers to display the spectra of numbers stations transmissions.

This is an area pioneered by John Fulford, WA4VPY. He uses a Cushman CE-15 spectrum analyzer and feeds it the intermediate frequency of his receiver. John has made several fascinating discoveries as a result of spending numerous hours monitoring number signals transmissions with this arrangement. For example, while monitoring four-digit Spanish transmissions on 6840 kHz, John discovered that a data burst is sent each 15 minutes with some transmissions. He has also determined that at least one other carrier is present on some frequencies when a numbers transmission is being made; the second carrier is so close in frequency to the numbers station that no heterodyne is heard on the channel. Further spectrum analysis has produced some evidence of subaudible tones being sent on some numbers transmissions (usually of the four-digit variety) and also of phase-shifting of the carrier in order to transmit data in binary format.

The discovery that multiple closely-matched carriers are present on a frequency carrying a numbers transmission has led to speculation that multiple sites may be used for some transmissions. As detailed in Chapter 3, it seems clear that multiple transmitter sites are used for some communications, such as those involving Air Force One and the Federal Emergency Management Agency (FEMA). The use of multiple transmitter sites would help explain such things as how certain numbers stations are simultaneously heard with local-strength signals in widely scattered locations such as New England, southern Florida, Colorado, and California. Future work with direction-finding antennas and spectrum analyzers will hopefully confirm or deny such speculation.

There are many other unresolved issues concerning numbers stations, and there will doubtlessly be additional new mysteries about them in the years ahead. The only safe statement that can be made about them is that numbers stations will continue to hold a powerful fascination for many SWLs.

2

Monitoring Mystery Networks

NOT ALL STRANGE CW ACTIVITY you can hear will be numbers stations. There are some CW nets using unusual call signs which can be heard exchanging cipher traffic with each other. These stations often use manually-keyed CW (often with an old "straight" key, not a "bug" or electronic keyer!) and the sending can be very sloppy. The operating procedure can be informal and somewhat disorganized. On the other hand, some of the nets have a very professional "sound" with good CW technique and formal check-in/check-out procedures, including the use of military "Z" codes. Strange CW markers have also been heard. Some of these are "normal" CQ, VVV, or QRA markers sent by stations with unusual call signs. Others are seemingly random groups of letters and numbers sent repeatedly.

Covert CW Networks

Let's take a closer look at a few of these unusual CW operations. While working on this book I received a tape of seemingly nonstop CW transmission. A monitor in England had frequently observed these very strange and very long messages and forwarded the tape in the hopes it would aid in identification of the activity.

In the comments accompanying the tape, he mentioned he decided one night to stick it out and try to hear the end of a message.

He listened one time for abut three hours and another time for about five hours, but both times the transmissions just stopped with no formal message ending or signdown. He pointed out that these signals reminded him of the Energizer bunny—they just keep going and going and going . . .

Listening to the tape revealed text groups of five characters each, and 44 different characters were noted in use. These were: letters A to Z; numbers 1 to 0; punctuation marks like the colon, comma, slant bar, hyphen; and special characters like AA (A with 2 dots over it), WA (accented A), UI (accented E) and UA (unknown equivalent).

These signals have been heard in Europe and England in the 2 to 6 MHz bands at night and 9 to 11 MHz bands during daylight hours. It will be interesting to see if U.S. monitors also hear these transmissions.

I find it rather strange that no one has ever copied a message heading or a message ending. The transmission just stops, and in the traffic I copied from the tape I could not see any particular significance to the last group of the text which was sent. Also, no one has ever copied any call signs associated with this activity. It seems as if a tape is turned on which plays for hours, then is turned off. Do you suppose this is a training operation?

Here is another strange CW activity. The oddball call sign is what caught my attention when I tuned into the signal. It was A2, so I started to copy. Here is what I heard:

A2, DE D1 Y42 AAA MSG NR 90834 BK
WMF ZPS VAH OGX RZJ K (heard 2214 UTC)
A2, DE D1 Y42 AAA MSG NR 90835 BK
GEL HOC IYR MJK TCE K (heard 2215)
A2, DE D1 Y42 AAA MSG NR 90836 BK
ODX IGC YLE RTF CHK K (heard 2216)
A2, DE D1 Y42 AAA MSG NR 90837 BK
DEN VIY HXZ UTR GAE K (heard 2217)
A2, DE D1 Y42 AAA MSG NR 90838 BK
XSO GWP DEC RKH PAG K (heard 2218)

The signal was very strong automatic Morse. No other stations were heard on the frequency which was 17479.8 kHz. Every now and then, I check for this traffic broadcast, but I have not encountered it again.

In the September, 1988 "Communications Confidential" column of ***Popular Communications*** magazine, I briefly described some unidentified CW and RTTY transmissions which I and others had copied from time to time. In order to learn more about this activity, increased monitoring had been undertaken resulting in approximately 160 RTTY messages collected along with a quantity of CW messages.

At the time I saw no evidence of two-way communications for any of the network stations. All transmissions seemed to be blind broadcasts. Seven callsigns were observed with two of them, YBU and JMS, seen on the RTTY broadcasts and the other five; PSN, ROL, BPA, SPK and WNY, appearing as addresses on the high-speed automatic Morse broadcasts. In comparing traffic flow to the various stations, it was readily apparent that a rather considerable amount of traffic was sent to station YBU at various dates within the monitoring period.

Tables 2-1 and 2-2 which follow the network composition as it was developed from the results of the 1987/1988 monitoring.

Table 2-1

CW TRANSMISSIONS

Frequency	*Calls*	*Day*	*Schedules (UTC)*	*Type of Traffic*
16457.7	PSN	MON	2250	5F
16473	ROL	TUE	1400	5F
16457.5	BPA	TUE	1530, 2300	5F & Svc Msg
16457.7	BPA	WED	1530, 1630, 2300	5F
16457.4	SPK	THU	1615	5F
	WNY		1900	Svc Msg
	PSN		2250	5F
16447	?	FRI	1525	5F
	SPK		1630	5F
	PSN		2255	5F
16457.4	BPA	SAT	1530	5F

NOTE: Most 5F traffic sent at very high speed.

Table 2-2

RTTY (7/425) TRANSMISSIONS

Frequency	*Calls*	*Day*	*Schedules (UTC)*	*Type of Traffic*
16456.3	YBU	MTWTFS	2200	Mostly 5L, some 5F msgs
16446.4	JMS	MTWTFS	2230	Mostly 5L, some 5F msgs

Network messages texts consisted of either 5F or 5L groups. In addition, a few QSL type messages were copied which concerned previous transmissions. Some Sunday monitoring was performed, but no Sunday schedules were discovered. All schedule times continued at the same UTC time when the U.S. changed from regular time to Daylight Savings Time in 1988.

The message headings were composed of five 5F groups as shown in the following examples:

Traffic to YBU –	11177 07386 21290 08636 01979
	11199 07386 00000 08637 00049
Traffic to JMS –	11177 00484 19117 07079 01909
	11177 00484 34133 06078 01199

A study of the headings revealed the following points:

First group:	Possible addressee indicators.
Second group:	Meaning undetermined.
Third group:	Either 00000 or 5 random numbers, meaning undetermined.
Fourth group:	Digits 1 & 2 show date, other 3 digits represent some type of serial number.
Fifth group:	Digits 1-4 show quantity of text groups in message plus 1. Reason for +1 not determined. Digit 5 is either a 1 or a 9. Meaning undetermined.

The last group of the 5L texts was found to break out through the use of a QWERTYUIOP = 1 to 0 decipherment into what may possibly be a cryptographic indicator. The letters A and/or Z were often seen in combination with the deciphered numbers. During the monitoring period, the only chatter noted was when "ALL TXT CFM" was seen following the transmission of several messages. Upon the completion of transmissions, "QRU QRU SK SK" was always sent for both the CW and RTTY modes.

The RTTY callup was a line of "46's" (not RY's) followed by a line with three repetitions of the callsign, number of messages upcoming, slant bar, total group count to be transmitted. The amount of traffic collected during the initial coverage was not considerable but nevertheless it was sufficient to yield some interesting details.

The results of this monitoring project certainly piqued my interest, and I decided that in the future, I would take another look at this activity. During the ensuing years, the CW broadcasts of this unidentified activity had evidently dried up as I and other monitors were no longer seeing those transmissions, but we were noting an increased number of the RTTY traffic broadcasts.

In late 1991, I decided to mount an extremely concentrated traffic collection with daily monitoring whenever possible, and this effort continued up into the later fall of 1993.

The expanded long term monitoring revealed a much larger network than that shown by the brief initial coverage which was previously described. Also, it was now pretty evident that RTTY seemed to be the exclusive mode for transmitting messages. While some of the same callsigns were again being seen, many additional callsigns were being discovered.

Although traffic has been seen being sent to a station on Mondays through Saturdays, this doe not mean that this is necessarily the case all those days of each an every week. After many hours of Sunday monitoring, I came to the conclusion that such schedules are not normally held but instead are probably arranged if world political or military events dictate that Sunday communications were required.

The main reference for the analysis of this network was "Traffic Analysis and the ZENDIAN Problem" by Lambros D. Callimahos. The book is a reprint of a formerly classified U.S. Government training text.

NETWORK SCHEDULES

Time UTC	*Recipient Callsign*	*Link Designator*	*Baud/Shift*	*Frequencies kHz*
1400	YBU	00148	75/500	16226.3/18803.2
1415	WFO	00125	75/500	14736
	MIG	00125	75/500	13382
1515	BPA	00116	75/500	14824/10424
1600	SPK	00168	50/500	21865.3/18844.2
1645	GMN	00119	75/500	20731.3/18126.8
1650	NDO*	00156	50/500	17488.2
1735	KRN	00178	75/500	18446.1/16446.2
1815	WFO	00125	75/500	14736
	MIG	00125	75/500	13382
1845	BPA	00116	75/500	19610.4/14812.4
1905	WNY	00139	75/500	20586.3/16446.2
1925	PSN	00126	75/500	16446.4/_____
2000	HZW	00117	50/500	20091/18196
2005	YBU	00148	75/500	19613/_____
2050	KAC	00128	75/500	18739.4/16891.2
2100	BAR	00135	50/500	18126.2/16135.2
2115	WFO	00125	75/500	14736
	MIG	00125	75/500	10841.7
2200	YBU	00148	75/500	20138.2/17478.2
2230	JMS	00127	75/500	16841.2/13625.3
2250	PSN	00126	75/500	23629.3/19955.3

**The NDO schedule is once a week on Wednesdays.*

The broadcasts for each callsign (except WFO/MIG) are initially transmitted on a primary frequency and then repeated on a secondary frequency. As indicated on the schedule chart, there are several frequencies as yet unrecovered. Plus there may also be some additional callsigns not yet detected.

The 1815 WFO/MIG, 1845 BPA, 1925 PSN and the 2005 YBU schedules do not appear to be held on a regular basis, but rather may be utilized for those days when traffic volume is very heavy. WFO frequently requests the transmit site to QSY to other frequencies in an apparent attempt to improve his signal reception. Here are the frequencies (in kHz) seen in use:

10231.4	12455	13627	16137	17501
10842	12536	14391	16232	17599
11106	12678	14462	16446	17611
11404	13014	14726	16888	18091
12004	13152	14968	16896	18448
12440	13446	15460	17493	18713
				19172

On March 24, 1992 at approximately 1800 UTC, two stations were intercepted on 16137 kHz using procedures like those used by stations on the target network. Station A told Station B to QSY to 24432 kHz. Station B was found on 24431.4 kHz tuning his transmitter, and he then began sending RTTY. The signal was very weak so good copy was not possible, but it looked as if the usual 4646's were being transmitted, and he then went into traffic. The signal was still very poor, so I could not get intelligible copy, but I left the receiver on the frequency. I dropped 16137 kHz to check on some other frequencies, and when I returned to 16137 kHz, I heard Station A giving a QSL for a message by repeating the message heading which was seen to be of the identical type in use on the target network. The repeated heading was as follows:

QSL 11177 00148 00000 24531 01731

This was followed by Station A giving the QSL time as QTR 1811 which corresponded to the UTC time at that moment.

I checked several of my receiving antennas and found one that brought in the signal of Station B much better. I was now able to obtain decent copy of Station B on 24431.4 kHz as he transmitted in RTTY giving a recapitulation of all the messages he had sent.

Here is what the traffic recapitulation looked like:

(in here) NR 132 GR 121

4M FOR 116 NR 098 GR 104
5M FOR 148 NR 530 GR 104
6M FOR 116 NR 099 GR 121
7M FOR 125 NR 432 GR 121
8M FOR 168 NR 053 GR 5
9M FOR 127 NR 150 GR 173
10M FOR 148 NR 531 GR 173
ALL QTC 10: FOR 135-1, FOR 127-2,
FOR 125-2, FOR 116-2, FOR 148-2, FOR 168-1

(Note: The "M" above is evidently an abbreviation for "message.")

You will no doubt spot that the trinomes following the "FOR" in the message recapitulation equate to the designators represented by the 2nd group in message headings but without the zeros.

Sample Callup When No Traffic Is Passed

4646464646464646464646464646464646464646464646464646464646464646464
JMS JMS JMS JMS JMS JMS JMS JMS JMS JMS JMS JMS JMS JMS JMS
4646464646464646464646464646464646464646464646464646464646464646464
JMS JMS JMS JMS JMS JMS JMS JMS JMS JMS JMS JMS JMS JMS JMS
4646464646464646464646464646464646464646464646464646464646464646464
JMS JMS JMS JMS JMS JMS JMS JMS JMS JMS JMS JMS JMS JMS JMS
4646464646464646464646464646464646464646464646464646464646464646464
JMS JMS JMS JMS JMS JMS JMS JMS JMS JMS JMS JMS JMS JMS JMS
4646464646464646464646464646464646464646464646464646464646464646464
JMS JMS JMS JMS JMS JMS JMS JMS JMS JMS JMS JMS JMS JMS JMS
4646464646464646464646464646464646464646464646464646464646464646464
JMS JMS JMS JMS JMS JMS JMS JMS JMS JMS JMS JMS JMS JMS JMS
4646464646464646464646464646464646464646464646464646464646464646464
JMS JMS JMS JMS JMS JMS JMS JMS JMS JMS JMS JMS JMS JMS JMS
4646464646464646464646464646464646464646464646464646464646464646464
JMS JMS JMS JMS JMS JMS JMS JMS JMS JMS JMS JMS JMS JMS JMS
4646464646464646464646464646464646464646464646464646464646464646464

JMS JMS JMS JMS JMS JMS JMS JMS JMS JMS JMS JMS JMS JMS JMS

QRU QRU SK SK

Typical Message of 5L Groups

11166 00178 4435 23010 00339
MMFJO OYSJH JAULV RDACI ODNML QHFZW MYSXR ZQBBN ZTSZR PYDUX ISHWM LFSPI SBHTQ LYLFG ERDCA KBBSI ODIHE WLWXV PVFOA MGZJW BIGZF MYFHA MSIDZ UOVJT SCAWT KTYEA OHTXC YACNN HDBVK QLJDR JVRXO UZOZO

Typical Message of 5F Groups

11177 00178 00000 25011 00451
69242 86607 75884 41521 52676 18250 97928 37067 40498 71479 93059 44251 11220 75032 99373 90046 19472 16538 14047 92317 68946 68222 78132 28097 14332 81869 34019 50181 70990 34919 79277 11569 79373 56941 58862 49878 69050 08088 96171 08733 05411 64048 91598 11111

QSL Traffic

In the listing of message headings I have indicated one of the types of traffic as being "QSL." These messages are always seen with 11199 as the first group of the heading and the text is only 1-5 (approximately) groups in length. The text groups look like message serial numbers as you will note in the examples.

TO GMN - 11199 00119 00000 04280 00029 26900
TO SPK - 11199 00168 00000 22080 00029 01400
TO WNY - 11199 00139 00000 12038 00029 03400
TO WFQ - 11199 00125 00000 10663 00099
70000 70100 70200 70300 70400 70500 70600 70700

When you scan through the headings for each callsign, I believe you will come to the same conclusion regarding the messages with 11199 as the first group of the heading.

Network Technical Notes and Operator Chatter

The technical notes are often preceeded by the word "TIKAS." The exact breakout of this term is not known, but it is readily apparent that these notes are technical in nature since they deal with frequency changes, schedule changes, frequency testing, establishment of special schedules during holiday periods and for upcoming urgent messages.

Both the technical notes and the operator chatter are in abbreviated English with many of these abbreviations being in a corrupt form. For example: VY TKS = "very thanks." However, the same abbreviation is seen correctly used as in VY FB = "very fine business."

While some Q-signals seem to be legitimate, there are many which evidently have meanings different from the normal meanings. QSF is one which definitely has a specialized definition. This particular Q-signal is used in the network in connection with TIKAS relating to holiday schedules. The international usage is "I have effected rescue and am proceeding to . . . base (with . . . persons injured requiring ambulance)."

For recognition purposes, some samples of both the technical notes and of operator chatter have been included.

On December 30, 1992 a technical note was sent which seemed to indicate a schedule arrangement for the period of New Year's Eve to January 3, 1993.

31/12	04/01 -	QSF 1
01/01	02/01 -	QSF 4
03/01	-	QSF 3

The above note was sent to HZW, SPK, KRN, WNY, KAC, and BPA. It was probably sent to others in the network also.

Other types of technical notes include these:

10/29/92 TO WNY – 1. QSO 19.00 QRG OLD
2. FM 1/11 QWK QDG TO NEXT QWK:
FM 00.00 TO 13.00 QSW 10439/7886/14834

FM 13.00 TO 24.00 QSW 16898/14834/19383

11/03/92 TO GMN – QWJ 14/12 QTA QWY 14/11
12/04/92 TO GMN – FM 7/12 QSO 4 WRK QTR 16.45
12/30/92 TO HZW – FM 1/1 QSO 5 QSW 21810/19612
02/01/93 TO GMN – OM PSE SEE TIKAS:
UR QTC QPP 29.01-HR 032 ES 31.01-NR 032
NR 101 COL QTC 31.01-NR 033 ES NEXT QTC QPP
WILL NR 034
02/12/93 TO WNY – NR 101 SA: QXU NR 032
05/. ./93 TO WFO – 1/6 TO NEXT QWK
FM 2300 TO 1200 QSY 10856/9369/7735
QSW 12455/10233/8109
FM 1200 TO 2300 QSY 10856/13372/15559
QSW 12455/14726/13448
08/24/93 TO BPA – FM 25/8 QSO 15.15 QSW 14824/10424
10/13/93 TO GMN – NR 101 SA: UR QSP ES RPT QTC NR 199 OM PSE QSL
12/06/93 TO WFO – QBG FM 06/12 TO NEXT QWK
2300 TO 1400 QSY 10856/7735/5102
QSW 10245/8108/4532
1400 TO 2300 QSY 10856/13372/15559
QSW 10243/14726/12440

To date only one link has been detected as having two-way communications. This is the WFO/MIG link. WFO sends the callup in FSK Morse while MIG sees to be using CW for the initial callup as well as for short messages of just a couple of groups. Otherwise the mode is RTTY for both stations.

Some of the abbreviations have been figured out. TXT is "text" while TX is "transmitter." Some others are obvious, some not.

Here are some samples of the WFO/MIG chatter exchanges:

HR VY QRN TX NO WRK PSE AS ALL
ALL TIME FB
AS 15
OK AS 15

On another date this was sent:

HR QSG NOT QBD
SRI PSE NR 935 AGN
WRK VY FB VY TKS

Personal chatter has been directed to the operator at NDO on several occasions as shown in the following:

5/27/92 – LTR NO, SA UR TLKF, 73, P.
11/18/92 – HR ALL FB, TMR HR OP-2, AS LTR, 73, P-OP-1
6/30/93 – OM AS TLF 03.00 GMT 01.07.93 P.
12/8/93 – OM, SA 158: HR SULTAN BORN DOTHER (3500/54), ALL FB I HR 3 YEAR. 73. P.

Here is my 1993 analysis of heading groups:

First group: To date only four different combinations have been seen as this group. They are: 11177, 11166, 11144, and 11199. The specific purpose of the first three designators has not been firmly determined. A few possibilities would be organizational components, security classifications, or perhaps categories of subject matter. The 11199 group however seems to be directly related to acknowledgment of receipt of a message. Thus, in the listing of headings, I have identified that traffic type as being QSL traffic.

Second group: Suspected as serving as a designator of a particular link. Each link seems to have its own unique second group. For example: The WFO?MIG link has 00125 as the 2nd group in the heading regardless of which one of the two stations was the originator of the particular message. By the way, the WFO/MIG link is the only one noted as being a two-way link. All of the other schedules do seem to be blind broadcasts with perhaps the response being transmitted at a later time.

Third group: Either five zeros or what seems to be five random digits. It is believed this group serves a cryptographic function in combination with the last group of the message text. It is also believed this group is included in the message group count. Refer to the comments for the fifth heading group.

Fourth group: The first two digits of the group represent the date of origination of the message and the last three digits are a consecutively numbered message serial number system for each recipient.

Fifth group: Digits 1 through 4 make up the group count of the message which is always the actual number of text groups plus one. It is suspected that the third group of the heading is counted as a group of text. If the supposition regarding the cryptographic nature of the third group is correct, it would seem to make sense that it be included in the total indicated for the group count. The numbers 1 or 9 are the usual ones seen in the final position of the fifth group. Traffic to MIG from WFO usually has the number 3 in that position. The significance of the use of just these three numbers remains to be determined.

Readers are invited to carefully look at the logged headings which follow and you will quickly recognize the various features described herein. You will no doubt notice a number of instances where identical texts were sent to several different stations. This fact shows up when comparing group counts and final text groups for pieces of traffic sent on or about the same date.

Regarding the message serial numbers in the fourth group, on January 1 each year the message number reverts to 001.

As mentioned earlier, the last group in the messages of 5L groups breaks out to digits with the inclusion sometimes of letters A and/or Z. The digits 1 to 0 equal the typewriter keyboard of QWERTYUIOP. It is suspected that this final text group, in conjunction with the third group of the message heading, are the cryptographic indicators.

To illustrate the equivalency, here are a few final text groups with their respective breakouts.

PTPWE = 95923; ZIIET = Z8835; OAUOO = 9A799

While some of the logged headings on file date back into 1991, I have tried to list the more recent ones for those addressees which usually have large volumes of messages. Others with lesser amounts indicate that the coverage was conducted over a greater time span.

When coverage on consecutive days turned up missing message numbers for a particular callsign, this was a pretty good indication that additional broadcasts or contacts were taking place during the period of roughly 2300-1300 UTC.

I do want to point out that I and others did not devote as much effort to that time frame as we did to that of the 1300-2300 UTC period.

Headings of Messages to BAR

Note: Periods (.) indicate missed or garbled characters.

		Heading Groups						*Text*
Day	*Date*	*1st*	*2nd*	*3rd*	*4th*	*5th*	*Tfc*	*Last Grp*
Mon	23 Mar 93	11177	00135	00000	21507	00031	5F	11111
		11177	00135	55608	23058	00989	5L	UZOPR
Fri	26 Mar 93	11166	00135	36871	26061	00809	5L	UEOWW
Wed	07 Apr 93	11144	00135	08435	07069	01949	5L	OAIPI
Tue	20 Apr 93	11166	00135	00000	20078	02901	5F	11111
Thu	22 Apr 93	11166	00135	00000	22081	00841	5F	11111
Fri	30 Apr 93	11166	00135	00000	30085	01211	5F	11111
Tue	11 May 93	11144	00135	51758	11090	00029	5F	71777
Thu	20 May 93	11177	00135	00000	19098	00871	5F	11111
Tue	25 May 93	11177	00135	78793	25105	0319.	5L	URZOW
		11177	00135	00000	25106	08671	5L	. . . . .
Wed	26 May 93	QRU						
Thu	27 May 93	QRU						
Mon	07 Jun 93	11166	00135	00000	07112	00561	5F	11111
Thu	24 Jun 93	11177	00135	48195	. . . . .	. . . . .	5L	. . . . .
Fri	25 Jun 93	11144	00135	11608	25118	00809	5L	UOWW
Thu	08 Jul 93	QRU						
Thu	22 Jul 93	QRU						
Thu	05 Aug 93	QRU						
Thu	12 Aug 93	QRU						

Mon	23 Aug 93	11166	00135	41452	231..	01939	.	.....
Thu	26 Aug 93	QRU						
Tue	14 Sep 93	HAD SKED, UNCOPY HERE, QRM						
Thu	18 Nov 93	QRU						
Mon	06 Dec 93	11166	00135	19805	06180	00789	5L	OEOWR

To BPA

		Heading Groups					*Text*	
Day	Date	1st	2nd	3rd	4th	5th	Tfc	Last Grp
Fri	27 Aug 93	11177	00116	58448	27162	01929	5L	UWIAP
		11177	00116	36721	27163	00299	5F	64154
		11177	00116	00000	27164	00561	5F	11111
Mon	30 Aug 93	11166	00116	66754	29165	00029	5F	17576
Thu	16 Sep 93	11177	00116	00000	15177	01381	5L	IRIZR
		11177	00116	60926	15176	00979	5L	IROPW
Mon	27 Sep 93	11177	00116	00000	26179	00031	5F	11111
Tue	28 Sep 93	11177	00116	13407	28180	02309	5L	UAUUW
		11177	00116	76007	28181	00469	5F	
		RERUN OF MSG NR 179						
Tue	05 Oct 93	11144	00116	00000	05189	03871	5L	OTZAW
		11144	00116	00000	04188	05241	5L	OTRUI
		11177	00116	00000	04186	00031	5F	11111
		11177	00116	77665	04187	00679	5L	OTOET
Tue	12 Oct 93	11166	00116	50990	11195	00839	5L	IIOAO
		11177	00116	08682	11196	02409	5L	IIUZW
Wed	13 Oct 93	11177	00116	00000	12197	01391	5L	IUIZE
Wed	20 Oct 93	11177	00116	00000	19204	02101	5L	IPUOW
		11199	00116	00000	20205	00029	QSL	01100
Fri	22 Oct 93	11177	00116	00000	20206	03201	5L	UOZIW
		11177	00116	00000	21207	03481	5L	UIZTA
Wed	27 Oct 93	11177	00116	00000	25210	00611	5F	11111
Fri	29 Oct 93	11177	00116	44920	28211	01159	5L	UAIIU
		11177	00116	47229	28212	00769	5L	UAOWZ
		11177	00116	25504	29213	00869	5L	UPUAW
Mon	01 Nov 93	11177	00116	84443	30216	00869	5L	ZOOAZ

Day	Date	Heading Groups 1st	2nd	3rd	4th	5th	Tfc	Text Last Grp
		11177	00116	97196	29215	01259	5L	UPIUU
Tue	02 Nov 93	QRU						
Thu	04 Nov 93	11177	00116	00000	03221	02151	5F	11111
		11177	00116	38072	03218	00449	5L	OZOTU
		11199	00116	00000	03219	00029	QSL	01300
		11177	00116	20822	03220	00649	5L	OZOEI
Mon	08 Nov 93	11177	00116	00000	07223	00031	5L	OZOEI
Tue	09 Nov 93	RERUN OF MSG NR 223						
Thu	11 Nov 93	11177	00116	25567	10227	00809	5L	IOOWW
		11166	00116	36198	09224	00539	5L	OPORO
		11199	00116	00000	10226	00039	5F	55555
		11177	00116	00000	10226	05141	5L	IORII
Mon	15 Nov 93	11177	00116	00000	15228	00031	5F	11111
		RERUN OF MSG NR 227						
Thu	25 Nov 93	11166	00116	88507	24236	01859	5L	UTIAU
		11177	00116	00000	23235	06031	5L	UZEOO
Tue	07 Dec 93	11177	00116	37441	06239	00279	5L	OEOUT
		11177	00116	00000	07240	01851	5L	OWIAU
Wed	08 Dec 93	11177	00116	94592	07241	01669	5L	OWIEZ
		RERUN OF MSG NR 240						
Fri	10 Dec 93	11177	00116	23590	09243	00739	5L	OPOWO

To GMN

Day	Date	Heading Groups 1st	2nd	3rd	4th	5th	Tfc	Text Last Grp
Wed	21 Apr 93	11199	00119	00000	21101	00029	QSL	09400
		11177	00119	10369	21102	00669	5F	11111
		11177	00119	00000	21103	00791	5F	11111
Thu	22 Apr 93	11199	00119	00000	22104	00029	QSL	09500
Fri	21 May 93	11199	00119	00000	21127	00029	QSL	10700
Mon	24 May 93	11177	00119	00000	23130	00031	5F	11111
Fri	28 May 93	11177	00119	68723	28134	00829	5F	82099
		11166	00119	51653	27133	00319	5F	67787

Mon	31 May 93	11177	00119	00000	30135	00031	5F	11111
Wed	02 Jun 93	11177	00119	64231	01237	03869	5L	60886
		11177	00119	93671	01136	00289	5L	OIOUR
Thu	03 Jun 93	11199	00119	00000	02138	00029	QSL	11200
Mon	07 Jun 93	11177	00119	00000	06141	00031	5F	11111
		11199	00119	00000	07142	00029	QSL	12200
Mon	14 Jun 93	11177	00119	00000	13148	00031	5F	11111
		11199	00119	00000	13147	00029	QSL	12200
Mon	21 Jun 93	11177	00119	51562	21151	00359	5F	32636
		11177	00119	00000	18149	02061	5F	11111
		11199	00119	00000	19150	00029	QSL	12700
Tue	13 Jul 93	11199	00119	00000	13164	00039	QSL	14400
Sat	31 Jul 93	11177	00119	71694	30184	00999	5F	14221
Wed	04 Aug 93	11177	00119	45585	03186	00759	5L	OZOWU
Mon	23 Aug 93	11177	00119	00000	22191	00031	5F	11111
Fri	24 Sep 93	11177	00119	00000	22191	00031	5F	81531
Tue	28 Sep 93	11177	00119	00000	26206	00031	5F	11111
Tue	05 Oct 93	11144	00119	36407	04211	02279	5F	45483
Wed	06 Oct 93	11144	00119	25854	05212	02119	5F	02149
Tue	12 Oct 93	11177	00119	16060	11215	03209	5F	54621
		11999	00119	00000	12216	00039	QSL	19901
Mon	18 Oct 93	11199	00119	00000	17226	00029	QSL	20600
		11199	00119	00000	17227	00029	QSL	20700
Wed	20 Oct 93	11177	00119	47794	19229	01009	5L	IPOPW
		11177	00119	00000	20230	03201	5L	UOZIW
Fri	29 Oct 93	11199	00119	00000	28237	00029	QSL	21900
		11177	00119	00000	29239	01621	5L	UPUIR
		11177	00119	00000	29239	01621	5F	11111
Sat	30 Oct 93	11199	00119	00000	30240	00029	QSL	22000
Wed	03 Nov 93	11177	00119	94892	03241	00459	5L	OZOTZ
		11199	00119	00000	03241	00459	QSL	22300
		11177	00119	00000	03243	02151	5F	11111
Wed	10 Nov 93	11166	00119	98957	09249	00649	5L	IOOEI
		11177	00119	00000	10250	05141	5L	IORII
		11177	00019	20301	10251	00699	5L	IOOEE

				Heading Groups				Text
Day	*Date*	*1st*	*2nd*	*3rd*	*4th*	*5th*	*Tfc*	*Last Grp*
Mon	15 Nov 93	11177	00119	00000	15255	00031	5F	11111
Thu	18 Nov 93	11199	00119	00000	17257	00029	QSL	23300
Mon	22 Nov 93	11177	00119	00000	21258	00031	5F	11111
		11177	00119	00000	22259	01271	5L	UUIUT
Tue	07 Dec 93	11177	00119	00000	07264	01851	5L	OWIAU
Wed	08 Dec 93	11177	00119	43573	02365	01909	5L	OA.AW
Thu	09 Dec 93	11177	00119	94436	09266	00789	5L	OPOWR
Fri	10 Dec 93	11199	00119	00000	10267	00029	QSL	23900
	11177	00119	96634	10268	00349	5L	IOOZI	
Sat	11 Dec 93	RERUNS OF MSG NR'S 267 AND 268.						

To HZW

				Heading Groups				Text
Day	*Date*	*1st*	*2nd*	*3rd*	*4th*	*5th*	*Tfc*	*Last Grp*
Tue	29 Dec 92	11177	00117	93719	28106	00469	5F	65939
		11177	00117	00000	29107	00441	5F	11111
		11177	00117	00000	28105	00881	5F	11111
Wed.	30 Dec 92	11177	00117	00000	30108	01141	5F	11111
		RERUNS OF MSG NR'S 106 AND 107						
Tue	19 Jan 93	11177	00117	00000	19003	05611	5F	11111
Sat	30 Jan 93	11177	00117	00000	30004	03651	5F	11111
Mon	01 Mar 93	11177	00117	00000	28006	00031	5F	11111
Tue	02 Mar 93	RERUN OF MSG NR 006						
Sat	06 Mar 93	11177	00117	47358	05007	00859	SF	06007
Mon	08 Mar 93	11177	00117	00000	07008	00031	5F	11111
Tue	09 Mar 93	11177	00117	90747	09009	01659	5F	21094
		RERUN OF MSG NR 008						
Wed	10 Mar 93	11177	00117	17685	10010	00819	5F	13192
		RERUNS OF MSG NR'S 007 AND 009						
Thu	11 Mar 93	RERUNS OF MSG NR'S 007 AND 010						
Mon	15 Mar 93	11166	00117	00000	15011	05011	5F	11111
Tue	16 Mar 93	RERUN OF MSG NR 011						
Mon	22 Mar 93	11177	00117	00000	21012	00031	5F	11111

Tue	23 Mar 93	RERUN OF MSG NR 012						
Mon	29 Mar 93	11177	00117	00000	29014	01201	5F	11111
		11177	00117	00000	29015	02781	5F	11111
Tue	30 Mar 93	11177	00117	00000	29016	01921	5F	11111
		RERUNS OF MSG NR'S 014 AND 015						
Wed	31 Mar 93	RERUN OF MSG NR 016						
Mon	12 Apr 93	11177	00117	00000	11018	00031	5F	11111
Sat	17 Apr 93	11144	00117	00000	16021	00521	5F	11111
		11166	00117	65296	16020	00029	5F	09446
Tue	20 Apr 93	11166	00117	00000	20022	02901	5F	11111
Wed	21 Apr 93	11177	00117	00000	21023	00791	5F	11111
		RERUN OF MSG NR 022						
Mon	31 May 93	11177	00117	00000	30040	00031	5F	11111
Sat	05 Jun 93	11177	00117	00000	05041	00031	5F	11111
Mon	21 Jun 93	QRU						
Tue	22 Jun 93	QRU						
Mon	05 Jul 93	11177	00117	00000	04049	00031	5F	11111
Tue	06 Jul 93	RERUN OF MSG NR 049						
Tue	20 Jul 93	11166	00117	00000	20051	01041	5F	11111
Wed	04 Aug 93	QRU						
Wed	25 Aug 93	QRU						
Thu	23 Sep 93	11144	00117	66501	23062	00779	5F	57739
		11177	00117	00000	23061	01021	5F	11111
Fri	24 Sep 93	RERUNS OF MSG NR'S 061 AND 062						
Tue	05 Oct 93	11166	00117	47940	05064	00659	5F	73377
		11177	00117	00000	04063	00031	5F	11111
Wed	06 Oct 93	RERUN OF MSG NR 064						
Wed	13 Oct 93	QRU						
Wed	20 Oct 93	11177	00117	60989	19068	00389	5F	61831
Wed	27 Oct 93	QRU						
Tue	02 Nov 93	11177	00117	00000	01072	00031	5F	11111
Mon	08 Nov 93	11177	00117	00000	07074	00031	5F	11111
Tue	09 Nov 93	11177	00117	42354	09075	00629	5F	53871
		RERUN OF MSG NR 074						
Wed	10 Nov 93	RERUN OF MSG NR 075						

		Heading Groups						*Text*
Day	*Date*	*1st*	*2nd*	*3rd*	*4th*	*5th*	*Tfc*	*Last Grp*
Mon	15 Nov 93	11177	00117	00000	15078	00031	5F	11111
		11144	00117	01352	14077	00029	5F	15991

To JMS

		Heading Groups						*Text*
Day	*Date*	*1st*	*2nd*	*3rd*	*4th*	*5th*	*Tfc*	*Last Grp*
Tue	08 Dec 92	11177	00127	00000	07445	02691	5F	11111
		11177	00127	39662	08446	0408.	5L	OATOR
Tue	15 Dec 92	11177	00127	00000	14450	01071	5L	ITIOT
		11177	00127	00000	15451	01331	5L	IRIZO
Thu	17 Dec 92	11166	00127	00000	17. . .	02591	5F	11111
Fri	18 Dec 92	11177	00127	00000	18455	00631	5F	11111
Fri	22 Jan 93	11144	00127	00000	22019	01131	5L	UUIIO
Thu	28 Jan 93	11177	00127	00000	28026	00771	5F	11111
		11177	00127	00000	28027	08141	5L	UAAII
Fri	29 Jan 93	RERUN OF MSG NR 026						
Mon	22 Feb 93	11177	00127	00000	22049	08321	5L	UUAUO
Wed	03 Mar 93	11166	00127	71236	03055	02889	5L	OZUAR
		11166	00127	00000	03057	00911	5F	11111
Thu	04 Mar 93	11166	00127	00000	04060	00861	5F	11111
Fri	19 Mar 93	11166	00127	49515	19072	01639	5L	IPIEO
Tue	30 Mar 93	11177	00127	00000	30089	08971	5L	ZOAPT
Tue	06 Apr 93	11166	00127	98286	05094	03549	5L	OEZRI
Wed	14 Apr 93	11166	00127	76938	14106	01659	5L	IRIEU
Tue	20 Apr 93	11166	00127	00000	20. . .	02901	5F	11111
Wed	21 Apr 93	11177	00127	00000	21115	02791	5F	UIUWE
Fri	30 Apr 93	11166	00127	00000	30131	01211	5F	11111
		11177	00127	00000	29127	00751	5F	11111
		11177	00127	00000	30128	06361	5L	ZOEZZ
Tue	25 May 93	QRU						
Tue	01 Jun 93	QRU						
Thu	03 Jun 93	QRU						
Tue	10 Aug 93	11177	00127	47542	10246	02509	5L	IOUTW
Tue	21 Sep 93	QRU						

Tue	28 Sep 93	QRU						
Tue	05 Oct 93	11144	00127	00000	05297	03761	5L	OTZWZ
		11177	00127	25342	04294	01049	5L	. . . . .
Tue	12 Oct 93	QRU						
Wed	03 Nov 93	11177	00127	42343	03341	06669	5L	OZEEZ
		11177	00127	00000	03342	02151	5F	11111
Thu	04 Nov 93	11177	00127	.4233	03341	06669	5F	. . . . .
		11177	00127	00000	03342	02151	5F	. . . . .
Mon	22 Nov 93	11177	00127	00000	22357	01461	5L	UUITZ
Tue	07 Dec 93	QRU						
Wed	08 Dec 93	11177	00127	00000	08. . .	01921	5L	. . . . .

To KAC

		Heading Groups						*Text*
Day	*Date*	*1st*	*2nd*	*3rd*	*4th*	*5th*	*Tfc*	*Last Grp*
Sat	09 Oct 93	11144	00128	00000	09236	04691	5L	OPTEE
		11177	00128	24599	09235	03219	5L	OPZIA
		RERUN OF MSG NR 234						
Wed	13 Oct 93	11166	00128	36511	14238	01659	5L	ITIEU
		11177	00128	57796	14237	00769	5L	ITOW
Mon	18 Oct 93	11166	00128	64772	18243	01439	5L	IAITO
		11177	00128	00000	17241	00031	5F	11111
Tue	19 Oct 93	11166	00128	90829	19244	02109	5L	IPUOW
		RERUNS OF MSG NR'S 243 AND 241						
		A NOTE SENT "QTC NR 242 GR 134 QTA"						
Wed	20 Oct 93	11177	00128	68463	20245	01279	5L	UOIUT
		11177	00128	38933	20246	00819	5L	UOOWA
		11177	00128	16133	20247	00989	5L	UOOPR
		11177	00128	77382	20248	02689	5L	UOUER
		RERUN OF MSG NR 244						
Thu	21 Oct 93	11177	00128	00000	21249	03061	5L	UIZOZ
		11177	00128	94341	21250	00519	5L	UIOTA
		RERUNS OF MSG NR'S 245, 246 247 AND 248						
Fri	22 Oct 93	11177	00128	94157	22251	00719	5L	UUOEA
		RERUNS OF MSG NR'S 249 AND 250						

Day	Date	1st	2nd	3rd	4th	5th	Tfc	Text Last Grp
		Heading Groups						
Thu	28 Oct 93	11177	00128	83671	28255	08149	5L	UAAII
Fri	29 Oct 93	11177	00128	00000	29256	01621	5F	11111
		RERUN OF MSG NR 255						
Sat	30 Oct 93	RERUN OF MSG NR 256						
Mon	01 Nov 93	11177	00128	00000	01257	03191	5F	11111
Tue	02 Nov 93	11177	00128	99139	02258	01169	5L	OUIIZ
		RERUN OF MSG NR 257						
Wed	03 Nov 93	11177	00128	30426	03259	02959	5L	OZUPU
		11177	00128	00000	03260	02151	5F	11111
		11177	00128	78714	03261	01399	5L	OZIZE
		RERUN OF MSG NR 258						
Thu	04 Nov 93	11177	00128	99274	04262	03609	5L	OTZRW
		11177	00128	82046	04263	01429	5L	OTI . P
		RERUNS OF MSG NR'S 259, 260 AND 261						
Fri	05 Nov 93	11177	00128	53181	05266	00729	5L	OROEP
Sat	06 Nov 93	11177	00128	32846	06267	01419	5L	OEIZA
		RERUN OF MSG NR 266						
Mon	08 Nov 93	11177	00128	00000	07268	00031	5F	11111
		RERUN OF MSG NR 267						
Tue	09 Nov 93	11177	00128	25963	09269	02949	5L	OPUPI
Wed	10 Nov 93	11177	00128	00000	10270	05851	5L	IORAU
		RERUN OF MSG NR 269						
Thu	11 Nov 93	11177	00128	00000	11271	04061	5F	11111
		RERUN OF MSG NR 270						
Mon	15 Nov 93	11177	00128	00000	15272	00031	5F	11111
Tue	16 Nov 93	11166	00128	35511	15273	03799	5L	IEZWE
		11177	00128	65973	16274	00729	5L	IEOEP
		RERUN OF MSG NR 272						
Wed	24 Nov 93	11166	00128	12298	23278	01759	5L	UZIWU
		11166	00128	29071	23279	00029	5L	GUIEZ
		11177	00128	00000	22770	01461	5L	UUITZ
Fri	26 Nov 93	11177	00128	82366	26281	06089	5L	UEEOR
Mon	06 Dec 93	11177	00128	62856	06288	01369	5L	OEIZZ

		11177	00128	09603	06289	02419	5L	OEIZA
		11177	00128	78094	06290	01639	5L	OEIEO
Tue	07 Dec 93	RERUN OF MSG NR'S 288, 289 AND 290						

To KRN

		Heading Group						Text
Day	*Date*	*1st*	*2nd*	*3rd*	*4th*	*5th*	*Tfc*	*Last Grp*
Sat	03 Apr 93	11177	00178	93338	02034	02319	5L	OZUUA
		11177	00178	99729	02035	03939	5L	OZZPO
Mon	05 Apr 93	11177	00178	00000	04036	00031	5F	11111
		RERUNS OF MSG NR'S 034 AND 035						
Wed	07 Apr 93	11166	00178	33844	07038	02099	5L	OWUOE
		11177	00178	52222	07039	06829	5L	OWEWP
		11177	00178	90090	06037	03539	5L	OEZRO
Thu	08 Apr 93	11177	00178	51379	08040	03319	5L	OAZUA
		RERUNS OF MSG NR'S 038 AND 039						
Fri	09 Apr 93	RERUN OF MSG NR 040						
Mon	12 Apr 93	11177	00178	00000	11041	00031	5F	11111
Tue	13 Apr 93	RERUN OF MSG NR 041						
Fri	16 Apr 93	11177	00178	00000	16043	00711	5F	11111
Sat	17 Apr 93	11144	00178	00000	16044	00521	5F	11111
		RERUN OF MSG NR 043						
Mon	19 Apr 93	RERUN OF MSG NR 044						
Tue	20 Apr 93	11166	00178	00000	20045	02901	5F	11111
Wed	21 Apr 93	11177	00178	00000	21046	00791	5F	11111
		RERUN OF MSG NR 045						
Thu	22 Apr 93	RERUN OF MSG NR 046						
Wed	05 May 93	11177	00178	41583	05052	00269	5L	OROUE
Sat	08 May 93	11166	00178	17927	07054	00539	5L	OWORO
Fri	21 May 93	11166	00178	57656	20062	04539	5L	UOTRO
Sat	22 May 93	11166	00178	00000	21063	02281	5F	11111
		11177	00178	13414	22064	03539	5L	UUZRO
Thu	03 Jun 93	11177	00178	91809	03066	03879	5L	OZZAT
Fri	04 Jun 93	RERUN OF MSG NR 066						
Thu	10 Jun 93	11166	00178	00000	09069	00561	5F	11111

		Heading Groups						Text
Day	Date	1st	2nd	3rd	4th	5th	Tfc	Last Grp
Tue	15 Jun 93	11177	00178	00000	13070	00031	5F	11111
Mon	21 Jun 93	11177	00178	67490	21075	00669	5L	UIOEZ
		11177	00178	00000	18074	02061	5F	11111
Tue	22 Jun 93	RERUN OF MSG NR 075						
Wed	23 Jun 93	11177	00178	54802	23077	01409	5L	UZIZW
		11177	00178	10294	23078	01509	5L	UZITW
		11177	00178	58678	23079	00899	5L	UZOAE
		11177	00178	97782	23080	01539	5L	UZIRO
Thu	24 Jun 93	11177	00178	. . . . .	23076	00479	5F	02723
		RERUNS OF MSG NR'S 077, 078, 079, AND 080						
Fri	25 Jun 93	11177	00178	11264	25081	00899	5L	UROAE
Sat	26 Jun 93	11177	00178	00000	25082	06351	5F	11111
		11177	00178	80968	26083	01539	5L	UEIRO
		11177	00178	00000	26084	00621	5F	11111
		RERUN OF MSG NR 081						
Wed	14 Jul 93	11177	00178	46284	13089	02339	5L	IZUZO
Thu	16 Sep 93	11177	00178	16148	16128	03639	5F	45357
Tue	09 Nov 93	11177	00178	34561	08143	03129	5F	16882
Fri	10 Dec 93	11177	00178	66009	09152	06529	5F	67940

To MIG

		Heading Groups						Text
Day	Date	1st	2nd	3rd	4th	5th	Tfc	Last Grp
Wed	04 Aug 93	11177	00125	93386	. . . 56	07639	5L	OTWEZ
Mon	13 Sep 93	11177	00125	10925	13225	00023	5L	VURYR
		11177	00125	13735	13226	00023	5L	. PZLD
		11177	00125	49351	13227	00023	5L	WTVKJ
Fri	17 Sep 93	11177	00125	42024	17245	00023	5L	VP . . P
Tue	21 Sep 93	11177	00125	51..47	21257	00023	5L	SNMJL
Tue	28 Sep 93	11177	00125	. . . . .	28285	00059	?	. . . . .
		11177	00125	. . . . .	28286	0050 .	5L	UPORO
Wed	06 Oct 93	11166	00125	8 .9 . .	06331	03833	5L	OEZAO
		11177	00125	82023	06332	01583	5L	OEIRR

		11177	00125	58 .07	06333	05413	5L	OERZA
Wed	20 Oct	11177	00125	51023	20415	00023	5F	67521
		11177	00125	86976	20416	00023	5L	OOWIG
Fri	22 Oct 93	11177	00125	08528	22432	00023	5L	XJPDT
		.....	00125	.....	224..	.....	5L	UUITW
Wed	27 Oct 93	11177	00125	39875	27460	00043	5L	REFJW
Fri	29 Oct 93	11177	00125	.2781	28464	00023	5L	MJMQQ
Mon	01 Nov 93	11177	00125	06885	01480	00033	5L	FAENY
Wed	03 Nov 93	11177	00125	63922	03497	00023	5F	81080
		11177	00125	36705	03499	00023	5L	ERDLG
		11177	00125	.....	03498	13153	5L	OZZIU
Thu	04 Nov 93	11177	00125	..552	04501	04589	5L	OTTRR
		11177	00125	55705	04502	09839	5L	.....
Fri	05 Nov 93	11177	00125	02602	055 .8	0120.	5L	ORIIW
Mon	08 Nov 93	11177	00125	.....	.....	.....	5L	OAUOA
Sat	13 Nov 93	11177	00125	...56	13552	00023	5F	59225
		11177	00125	..480	13553	0005 .	5F	.....
Mon	15 Nov 93	11177	00125	36579	15561	00023	5L	YAKAI
Wed	24 Nov 93	11177	00125	07106	24615	00023	5L	XRETD

To NDO

				Heading Groups				*Text*
Day	*Date*	*1st*	*2nd*	*3rd*	*4th*	*5th*	*Tfc*	*Last Grp*
Wed	21 Oct 92	11199	00156	00000	20022	00029	QSL	00500
Wed	02 Dec 92	11199	00156	00000	01024	00029	QSL	00700
Wed	03 Mar 93	11199	00156	00000	02001	00029	QSL	00200
Wed	02 Jun 93	QRU						
Wed	09 Jun 93	QRU						
Wed	30 Jun 93	QRU						
Wed	04 Aug 93	QRU						
Wed	25 Aug 93	11166	00156	02850	25004	01829	..	.....
Wed	22 Sep 93	QRU						
Wed	13 Oct 93	QRU						
Wed	20 Oct 93	QRU						

Day	Date	Heading Groups 1st	2nd	3rd	4th	5th	Tfc	Text Last Grp
Wed	27 Oct 93	QRU						
Wed	10 Nov 93	QRU						
Wed	08 Dec 93	QRU nut sent note (birth announcement)						

The broadcasts to NDO have only been observed on Wednesdays, and then they are usually just the callup/QRU type. From May, 1992 to February, 1993, only two messages were intercepted going to NDO. based on this very scanty data, a "guesstimate" is that NDO receives one message each month and sends one message each month. Note the heading groups above. Message 022 was sent in October and it looked like it was a QSL for a message 005. Message 024 was sent to NDO in December and it appeared to be a QSL for a message 007. Thus, it may very well be possible that message 023 was sent in November to QSL a message 006.

While there is no way to confirm it, NDO may be the new (but not yet officially in operation) Russian embassy in Washington, DC.

To PSN

Day	Date	Heading Groups 1st	2nd	3rd	4th	5th	Tfc	Text Last Grp
Fri	22 Jan 93	11177	00126	10767	22004	00699	5F	09757
		11177	00126	47286	22005	00759	5F	13034
Mon	25 Jan 93	11177	00126	00000	25006	00451	5F	11111
Wed	27 Jan 93	11177	00126	05776	27007	00959	5F	55308
		11177	00126	00000	27008	04091	5F	11111
Sat	30 Jan 93	11177	00126	00000	300 . .	03651	5F	11111
Thu	04 Feb 93	11177	00126	45437	04012	00409	5F	59817
Thu	11 Feb 93	11177	00126	66286	11013	00219	5F	11637
Tue	16 Feb 93	11166	00126	00000	16014	01731	5F	11111
Wed	17 Feb 93	11177	00126	00000	17015	01241	5F	11111
Fri	26 Feb 93	11177	00126	00000	26019	03251	5F	11111
Mon	15 Mar 93	11166	00126	00000	15022	05011	5F	11111
		11177	00126	00000	15023	00591	5F	11111
Mon	22 Mar 93	11177	00126	00000	21024	00031	5F	11111

Mon	29 Mar 93	11177	00126	00000	29027	02781	5F	11111
		11177	00126	00000	29028	01921	5F	11111
Tue	30 Mar 93	11177	00126	00000	30029	00031	5F	11111
Wed	14 Apr 93	11177	00126	00000	14033	01671	5F	11111
Tue	20 Apr 93	11166	00126	00000	20 . . .	02901	5F	11111
Wed	21 Apr 93	11177	00126	00000	21308	00791	5F	11111
Tue	18 May 93	11177	00126	00000	18050	03621	5F	11111
Fri	02 Jul 93	11177	00126	98903	0	00279	5F	. . . . .
Mon	20 Sep 93	11177	00126	31125	18082	00959	5F	57153
Thu	28 Oct 93	11144	00126	296 . .	28095	. . . . .	5F	. . . . .
		11166	00126	. . . . .	2809 .	. . . . .	5F	. . . . .
		11177	00126	00000	25091	00611	5F	11111
		THREE MORE MSGS SENT BUT UNCOPY HERE, BAD QRN						
Wed	03 Nov 93	11177	00126	47945	03098	01319	5F	75725
		11177	00126	00000	03099	02151	5F	11111
Fri	10 Dec 93	BELIEVE VERY SHORT MSG PASSED BUT UNCOPY HERE						

To SPK

				Heading Groups				*Text*
Day	*Date*	*1st*	*2nd*	*3rd*	*4th*	*5th*	*Tfc*	*Last Grp*
Thu	22 Apr 93	11199	00168	00000	22114	00029	QSL	01800
Fri	23 Apr 93	RERUN OF MSG NR 114						
Tue	04 May 93	11177	00168	00000	30124	00921	5F	11111
Mon	24 May 93	11177	00168	00000	23410	00031	5F	11111
Fri	16 Jul 93	11177	00168	78809	15181	04289	5F	98140
Sat	17 Jul 93	11177	00168	00000	16182	01211	5F	11111
		11177	00168	00000	17183	04081	5F	11111
Tue	27 Jul 93	11177	00168	00000	25187	00031	5F	11111
Sat	31 Jul 93	QRU						
Fri	06 Aug 93	11177	00168	93097	06193	02959	5F	49727
Sat	07 Aug 93	11177	00168	48391	06194	05049	5F	08353
Fri	27 Aug 93	11199	00168	00000	27203	00039	QSL	02900
		11177	00168	00000	27204	00561	5F	11111
		11177	00168	00000	26202	00381	5F	11111
Thu	16 Sep 93	11177	00168	29059	16230	01009	5F	59151

Day	Date	Heading Groups 1st	2nd	3rd	4th	5th	Tfc	Text Last Grp
Mon	27 Sep 93	11177	00168	00000	26237	00031	5F	11111
Tue	05 Oct 93	11144	00168	16679	04243	02769	5F	27811
		11177	00168	47345	01239	04879	5F	01403
		11177	00168	00000	04242	00031	5F	11111
Wed	13 Oct 93	11177	00168	90603	13252	01019	5F	36009
Mon	18 Oct 93	11166	00168	40113	16256	08659	5F	49992
		11177	00168	00000	17257	00031	5F	11111
		11144	00168	83469	16255	03029	5F	42261
		11199	00168	00000	16254	00029	QSL	04300
Wed	20 Oct 93	11177	00168	63240	20255	01119	5F	98521
		11177	00168	11418	20259	01059	5F	75423
		11177	00168	38805	20260	01339	5F	71289
Thu	28 Oct 93	11177	00168	00000	26266	01841	5F	11111
		11177	00168	30608	26267	00379	5F	18671
Fri	29 Oct 93	11166	00168	38476	28268	00029	5F	44709
		11177	00168	00000	29269	01621	5F	11111
Sat	30 Oct 93	11199	00168	00000	30270	00029	QSL	04400
		11199	00168	00000	30271	00029	QSL	04500
Wed	03 Nov 93	11144	00168	77848	03278	00689	5F	95271
		11166	00168	03638	02276	00339	5F	01780
		11177	00168	03151	02275	00589	5F	62784
		11199	00168	00000	03277	00029	QSL	)$*))
		11177	00168	83294	03279	01079	5F	91991
		11199	00168	00000	03280	00029	QSL	04900
Mon	08 Nov 93	11177	00168	00000	03280	00029	QSL	04900
Mon	15 Nov 93	11199	00168	00000	14291	00029	QSL	05200
		11199	00168	00000	15292	00029	QSL	05300
		11177	00168	00000	15293	00031	5F	11111
		11199	00168	00000	15294	00039	QSL	05500
Thu	25 Nov 93	11166	00168	41870	25307	000249	5F	97241
Fri	26 Nov 93	11177	00168	16428	26308	01399	5F	49762
		RERUN OF MSG NR 307						
Sat	04 Dec 93	11177	00168	39611	02315	00569	5F	19415

		11177	00168	66211	03316	00869	5F	15645
Wed	08 Dec 93	11177	00168	98166	07321	01269	5F	92709
		11177	00168	42403	08322	01029	5F	82947

To WFO

				Heading Groups				*Text*
Day	*Date*	*1st*	*2nd*	*3rd*	*4th*	*5th*	*Tfc*	*Last Grp*
Tue	09 Nov 93	11199	00125	00000	09475	00059	QSL	52900
		11177	00125	00000	09476	00531	5L	OPORO
		11177	00125	51372	09477	00369	5L	OPOZZ
		11177	00125	67172	09478	12269	5L	OPUUZ
		11177	00125	02481	09479	00759	5L	OPOWZ
		11177	00125	61604	09480	07139	5L	OPWIO
		11177	00125	42581	09481	06049	5L	OPEOI
Wed	10 Nov 93	11177	00125	07139	10488	00539	5L	IOORO
		11177	00125	71691	10489	01199	5L	IOIIE
		11177	00125	90588	10490	08759	5L	IOAWU
Thu	11 Nov 93	11199	00125	00000	11491	00049	QSL	54000
		11177	00125	00000	11492	00501	5L	IIOTW
		11177	00125	00000	11493	01751	5L	IIOWI
		11177	00125	39887	11494	00749	5L	IIOWI
		11177	00125	00000	11495	02621	5L	IIURP
		11177	00125	00000	11496	04061	5F	11111
Fri	12 Nov 93	11177	00125	37216	12505	05419	5L	IURZA
		11177	00125	62833	12506	02219	5L	IUUIA
Sat	13 Nov 93	11177	00125	00000	12507	03561	5F	11111
		11177	00125	00000	13508	03771	5L	IZZWT
		11177	00125	00000	13509	02261	5L	IZUUZ
		11177	00125	00000	13510	04721	5L	IZTEP
		11177	00125	87871	13511	03369	5L	IZZZZ
		11177	00125	88945	13512	04779	5L	IZTWT
		11177	00125	00000	13513	30191	5L	. . . . .
Mon	15 Nov 93	11199	00125	00000	15516	00069	QSL	55900
		11177	00125	00000	15517	00741	5L	IROWI
		11177	00125	80351	15518	01019	5L	IROPA

Day	Date	*Heading Groups* 1st	2nd	3rd	4th	5th	Tfc	*Text* Last Grp
		11177	00125	57962	15519	01859	5L	IRIAU
Wed	17 Nov 93	11199	00125	00000	16532	00039	5F	55555
		11199	00125	00000	17533	00079	QSL	57400
		11177	00125	00000	17534	0.	5L	. . . . .
		11177	00125	31586	17535	00539	5L	IWORO
		11177	00125	00000	17536	02541	5L	IWURI
		11177	00125	00000	17537	00531	5L	IWORO
		11177	00125	32448	17538	04799	5L	IWTWE
		11177	00125	32965	17539	00969	5L	IWOPZ
		11177	00125	15788	17540	01469	5L	IWITZ
Fri	19 Nov 93	11177	00125	55298	19560	02219	5L	IPUIA
Thu	25 Nov 93	11177	00125	00000	25594	01151	5L	URIIU
		11177	00125	74058	25595	13259	5L	URZUU
Mon	06 Dec 93	11177	00125	00000	06639	00501	5L	OEOTW
		11177	00125	14779	06640	00499	5L	OEOTE
Tue	07 Dec 93	11177	00125	97553	07650	11209	5L	OWIIW
Fri	10 Dec 93	11199	00125	00000	10663	00099	QSL	70700
		11177	00125	61879	10665	01499	5L	IOITE
		11177	00125	34517	10666	02549	5L	IOURI
Mon	13 Dec 93	11177	00125	06758	13671	01669	5L	IZIEZ
		11177	00125	00000	13672	00541	5L	IZORI

To WNY

Day	Date	*Heading Groups* 1st	2nd	3rd	4th	5th	Tfc	*Text* Last Grp
Sat	03 Apr 93	11199	00139	00000	03086	00029	QSL	07600
Sat	10 Apr 93	11199	00139	00000	10089	00029	QSL	07800
Wed	14 Apr 93	11177	00139	00000	14091	01671	5F	11111
Fri	16 Apr 93	11177	00139	00000	16093	00711	5F	11111
Sat	17 Apr 93	11199	00139	00000	17095	00039	QSL	08500
Mon	19 Apr 93	11177	00139	93265	19096	01339	5L	IPIZO
Tue	20 Apr 93	11166	00139	00000	20098	02901	5F	11111

		11177	00139	00000	20097	04119	5L	UOTOA
Wed	21 Apr 93	11199	00139	00000	21099	00029	QSL	08600
		11177	00139	00000	21100	00791	5F	11111
Sat	24 Apr 93	11199	00139	00000	24104	00029	QSL	09200
Mon	24 May 93	11177	00139	00000	23132	00031	5F	11111
Wed	09 Jun 93	11199	00139	00000	09145	00029	QSL	14300
Thu	17 Jun 93	11199	00139	00000	17152	00029	QSL	14700
Mon	28 Jun 93	11199	00139	00000	28163	00029	QSL	16000
Tue	29 Jun 93	11199	00139	00000	29164	00039	5F	955555
Tue	24 Aug 93	11199	00139	16030	24207	00029	QSL	19300
		11177	00139	77969	24208	00409	5L	UTOZW
Fri	24 Sep 93	11199	00139	00000	24225	00039	QSL	21100
Mon	27 Sep 93	11177	00139	00000	26277	00031	5F	11111
Wed	06 Oct 93	11199	00139	00000	06239	00029	QSL	21900
Wed	13 Oct 93	11199	00139	00000	13245	00029	QSL	22300
Wed	20 Oct 93	11199	00139	00000	20251	00039	5F	55555
		11199	00139	00000	19250	00049	QSL	230000
Thu	28 Oct 93	11199	00139	00000	28257	00049	QSL	23800
		11177	00139	68518	28258	01439	5L	UAITO
Fri	29 Oct 93	11177	00139	00000	29259	01621	5F	11111
Sat	30 Oct 93	RERUN OF MSG NR 259						
Tue	02 Nov 93	11199	00139	00000	02262	00039	QSL	24100
Sat	06 Nov 93	11199	00139	00000	06268	00029	QSL	24600
Mon	08 Nov 93	11177	00139	00000	07269	00031	5F	11111
		11199	00139	00000	08270	00029	QSL	24700
Wed	10 Nov 93	11177	00139	65139	10272	00529	5L	IOOTP
Fri	19 Nov 93	11199	00139	00000	19281	00029	QSL	25700
Fri	26 Nov 93	11199	00139	00000	26289	00039	QSL	26400
		11199	00139	00000	25287	00029	5F	15700
		11199	00139	00000	25288	00039	5F	15711
Tue	07 Dec 93	11199	00139	00000	06297	00039	QSL	27100
Fri	10 Dec 93	11177	00139	24660	10299	00619	5L	IOORA

To YBU

Day	Date	Heading Groups 1st	2nd	3rd	4th	5th	Tfc	Text Last Grp
Mon	15 Nov 93	11177	00148	66609	15081	01189	5L	IRIIR
Wed	17 Nov 93	11177	00148	38152	16088	01289	5F	16359

Day	Date	Heading Groups 1st	2nd	3rd	4th	5th	Tfc	Text Last Grp
		11177	00148	26586	16089	01929	5F	73339
		11177	00148	23896	16090	01029	5F	17576
		11177	00148	16425	16091	05159	5L	IERIU
		11177	00148	47259	16092	02959	5L	IEUPU
		11177	00148	95161	16093	03169	5L	IEZIZ
		11177	00148	95244	16094	01809	5L	IEIWW
		11177	00148	47008	16095	01369	5L	IEIZZ
		11177	00148	94387	17096	04809	5L	IWTWW
		11177	00148	79012	17097	02599	5L	IWURE
		. . . . .	00148	. . . . .	17098	. . . . .	5L	IWORA
		11177	00148	85827	17099	01269	5L	IWIUZ
Mon	22 Nov 93	11177	00148	96659	22115	04849	5L	UUTAI
Thu	25 Nov 93	11177	00148	05107	25127	01089	5L	URIOR
		11177	00148	15652	25128	01059	5L	URIOU
		11177	00148	81137	25133	13099	5L	URZOE
		RERUNS OF MSG NR'S 127 AND 128						
Fri	10 Dec 93	11177	00148	93350	10174	00459	5L	IOOTU
		11177	00148	98989	10175	00359	5F	43939
Wed	03 Nov 93	11177	00148	00000	03042	02151	5F	11111
Thu	04 Nov 93	11177	00148	92415	04043	03449	5L	OTZTI
Fri	05 Nov 93	11166	00148	13050	05047	01059	5L	ORIOU
		11177	00148	18237	05044	01589	5L	ORIRR
		11177	00148	00000	05045	01411	5L	OIIZP
		11177	00148	41698	05046	00369	5L	ORO77
		11177	00148	74891	05048	01509	5L	ORITW
		11166	00148	40382	05050	00359	5L	OROZU
		11177	00148	53329	05049	01939	5L	ORIPO
Sat	06 Nov 93	QRU						

Mon	08 Nov 93	11177	00148	00000	07051	00031	5F	11111
Tue	09 Nov 93	11177	00148	24204	09052	02869	5L	OPUAZ
		11177	00148	98529	09053	12039	5L	. . . . .
		11177	00148	95513	09055	06069	5L	OPEPZ
		11177	00148	32305	09056	06799	5L	. . . . .
Wed	10 Nov 93	11177	00148	15709	10062	00459	5F	11759
		11177	00148	30785	10063	00739	5L	IOOWO
Thu	11 Nov 93	11177	00148	20825	11065	01989	5L	IIZRR
		11177	00148	94419	11064	03589	5L	IIIPR
		11177	00148	53979	1106 .	07099	5L	. . . . .
Fri	10 Dec 93	11177	00148	18214	10178	00589	5L	IOORR
		11177	00148	66504	10179	00359	5F	45302
		11177	00148	24858	10180	00469	5L	IOOTZ
		11177	00148	60637	10177	02559	5L	IOURU
		11177	00148	81295	10181	04929	5L	IOTAP
		11177	00148	57952	10182	01949	5L	IOIPO
Mon	13 Dec 93	11166	00148	27326	13186	00459	5L	IZOTR
		11166	00148	09022	13187	00569	5L	IZORZ
		11177	00148	94634	13184	02539	5L	IZURO
		11177	00148	80301	13185	01879	5L	IZIAT
		11177	00148	59006	13188	00539	5L	IZORO

Miscellaneous Reported Frequencies

UTC	*kHz*	*Callsign*	*Baud*	*Comments*
0313	13988	?	75	First heading group 11177
0907	14605	VKX	75	
1032	10945	SCJ	50	
1100	18225	VKX	75	
1125	13790.6	SFO	50	
1215	18128	BAR	50	
1230	12753	?	50	PSE QSL QRU SK
1245	16133.2	?	50	First heading group 11166
	19350.4	?	75	
1259	19101.7	DMA	50	
1332	6310	OTD	75	

1341	18342	?	75	First heading group 11177
1350	20233.2	?	50	Five letter and five-figure groups
1425	12237	KUL	75	
	10584	?	75	
1437	14377	?	75	
1455	17450	?	75	11177 20065 42394 09101 01649

UTC	*kHz*	*Callsign*	*Baud*	*Comments*
1503	18805	?	75	First heading group 11177
	14391	?	75	
1515	17444	?	75	Five-figure groups
1517	20467	?	75	
	22185.3	?	75	
1520	8166	LMS	75	
1540	13385.9	?	75	QRU QRU SK SK
1630	16327	YBU?	75	
1650	18356	?	75	Five-letter text

UTC	*kHz*	*Callsign*	*Baud*	*Comments*
1738	13833	?	75	Five-letter text
1753	13541	JMS	75	
1825	16138	YBU	75	
	19168	?	75	
1858	16300	?	75	
2044	16238	?	50	Five-figure groups
	16242	?	50	
2110	11540	CMU	50	

According to information in the August 1990 issue of ***Monitoring Times,*** the Morse code agent broadcasts that end their transmissions with either three or five cut zeros were reportedly KGB sponsored. Such broadcasts to Russian agents in the western hemisphere were said to be transmitted from the GRU communications facility located at Guineo, Cuba, which is about 15 miles SW of Havana.

The January, 1993 issue of ***MILITARY*** magazine carried an interesting item revealing Cuba and Russia had signed a new trade and economic agreement in November 1992 which included the continued operation of the large Russian SIGINT collection site which is located at Lourdes, Cuba, approximately 60 miles south of Havana. This installation was constructed by the Soviet Union in 1974 and is targeted against various U.S. communications. The agreement must certainly have also covered the continued use of the GRU transmissions facilities located near Havana.

On March 20, 1993 at 2113 UTC I detected what looked like an operator boo-boo. I think what happened was that the operator patched the tape of a Russian agent broadcast to the wrong transmitter, the one tuned up on 14735 kHz which had been brought for the upcoming 2115 WFO schedule. What came up on the frequency was "824 TTTTT 824 TTTTT 824 TT" and it was abruptly terminated. A moment later, the transmitter was keyed with the WFO callup. Other transmission mistakes seen included a message tape being run backwards. After a brief interval the transmission stopped and the tape was reversed, now with the message text being printed out correctly.

On November 23, 1992 the callup should have been for the WNY broadcast. Instead what was transmitted was PSN PSN PSN 1/61. The sequence was repeated the usual three times followed by a line of 46's etc. After only a few repetitions of the callup sequence, several minutes of "46's" were sent followed by the "WNY WNY WNY 1/3" sequence. There was a definite value to the above for analysis purposes because it indicated that callsign had last been heard during the coverage of the network in 1987/1988.

Another boo-boo took place on October 9, 1993 on the 2000 UTC schedule for traffic to HZW. The callup indicated "HZW HZW HZW 1/3" along with the "46's" sequence. However, the traffic was not transmitted after the callup.

I do wonder about a schedule to callsign RŌL. This station was receiving traffic during the 1987/1988 coverage but it was not observed during the 1991-1993 monitoring.

In reviewing the notes I was compiling regarding network operational characteristics, it was obvious that I generally experienced the best reception when I used the longwire antenna aimed toward Cuba. I also was aware that I usually heard most schedules at about the same, strong, signal strength level.

I recalled when TASS (the Soviet official news agency) was sending press on HF RTTY from Cuba to North and South America, and such material was believed to have been first passed from Moscow to Cuba via satellite. In "Soviet Signals Intelligence (SIGINT): Intercepting Satellite Communications" by Professor Desmond Ball of Australia, there was a mention of a Cuba/Moscow Molniya 1 satellite link which was used to transmit collected COMINT to Russia.

I wondered if that method might also be the one used to relay traffic of the network I was studying. This seemed to be a distinct possibility because I had only heard one instance of traffic suspected of coming from Moscow via HF RTTY for Cuba and that was the March 24, 1992 incident previously described. Perhaps on that particular date the satellite system was out of service?

On December 31, 1993 at approximately 1426 UTC, a sequence was heard in FSK Morse on 16873.7 kHz which sounded like "ETFNJX TKAGAS." This was sent until 1430, at which time it was discontinued. On that same date during the broadcast to SPK at 1600 on the priamry frequency of 21865.3 kHz, the broadcast was stopped and the transmitter keyed in FSK Morse and sent "ETFNJX TKAGAS." This continued for a few minutes and then ceased, and the traffic for SPK resumed. This FSK Morse sequence could also be "ETFNJZ QAG AS" or "E/NJX" or "AF NJX." It is sent in a "run together" fashion, making a correct determination of the intended sequence difficult. This type of transmission has been heard at other times on frequencies used for broadcasts to network stations. It would certainly appear, therefore, that there is a definite relationship between this FSK Morse calling and the operations of the Russian diplomatic network.

Similar Transmissions

In late 1992 and early 1993 additional RTTY activity was copied similar to that on the large network described earlier. One of the links was from an unlocated/unidentified transmitter sending to callsign VKX. The other schedule showed messages from station RCF. Callsign references had RCF listed as the Ministry of Foreign Affairs, Moscow. There was a VKX listed as Canberra, Australia, but it was very much in doubt that was the station in question.

Transmissions to VKX were heard weaker on the U.S. east coast than the signals to the large network but were stronger than the signals of RCF. However, in Europe the broadcasts to VKX were heard weaker than the RCF signals.

On a number of occasions, I had overheard hams complaining to each other about RTTY interference they were experiencing on 18126 kHz. This particular RTTY transmission was the one directed to callsign GMN of the large network. In April, 1993 I decided to write to the FCC and advise them of the complaints I had overheard. I hope this would perhaps help out in this matter and might also reveal the affiliation of the activity causing the interference. In the FCC reply to my letter, they indicated the transmissions were those of a station of the former Soviet Union and that station was authorized to use frequencies in the band on a primary basis. Based on the above plus other factors mentioned previously, I was now convinced that the RTTY communications described in this chapter were related and Russian sponsored.

Miscellaneous Schedules

Time UTC	*Recipient Callsign*	*Link Designator*	*Baud/Shift*	*Frequencies kHz*
1435	CQ DE RCF	Not Used	75/500	16201//16228//18450 19649
1500	VKX	00166	75/500	14605/12167.3

NOTE: RCF could frequently be heard on all four of the frequencies shown above. However, these frequencies have since changed and RCF is no longer heard.

When the first draft of this chapter was prepared, the secondary frequency for the VKX broadcast had not been detected. However, periodic checking was continued and the secondary frequency was finally found on December 29, 1993. There were a couple of more changes in the secondary frequency and in February, 1994 it was found to be 12178.4 kHz.

Many different addressee indicators have been seen on the messages transmitted by RCF. On the 1435 UTC schedule, these indicators were observed:

JUA RPO UDZ21 RKM BXL RXX FRU EWZ42 URO RMM RZJ BLA RCX81
BNV ACD FQX FJN UDZ27 TRP UXG CAZ RKG EWZ40 WNK KDN

On a 1640 UTC schedule on 14370 kHz these indicators were observed:

GLK DPL OBX PTF WKL KUA RSS LSG

On an unidentified schedule on 19649 kHz all of the 1435 UTC indicators were seen plus DSG KDN NMZ PTF.

In addition to RCF traffic, European monitors saw broadcasts to callsigns OTD, SCJ, CMU, DMA, YBU, RCA, and LMS.

Traffic sent by RCF often carries the phrase FOR ALL which indicates the message is for all indicated addressees.

Technical Notes

Here is a technical note sent to callsign VKX on December 4, 1992:

QSO 11.00 QRG 16840 QDD SA NEW QRG HR AS LTR

QOO has a specialized use because the normal definition is "I can send on any working frequency." RCF uses this Q-signal in itemizing message numbers with their corresponding group counts as shown in the following examples which also illustrate TIKAS messages.

CQ CQ CQ DE RCR CQ CQ CQ DE RCF
FOR FQX CAZ RQO BLA NR 303
FOR UDZ21 NR 202 101
FOR FJN EWZ42 NR 202

QOO NR 101 GR 203 NR 202 GR 88 NR 303 GR 638
TIKAS
- - - - -
FOR RKG ES FRU AFTER QKY QRZ

CQ CQ CQ DE RCF CQ CQ CQ DE RCF
FOR JUA RXX RZJ FQX UDZ21 CAZ RPO FRU BLA FJN RKG BXL URO
RMM BNV EWZ42 NR 146 48
FOR ACD EWZ40 TRP RCX81 UDZ27 UXG KDN NR 147 49 50
FOR RKM NR 48
QOO NR 146 GR 103 NR 147 GR 88 NR 148 GR 90
NR 49 GR 46 NR 50 GR 83
TIKAS
- - - - -
FOR ALL
1,2,3,9,10.05.92 WRK QSF 3
4,11.05.92 WRK QSF 2

The repetition was carried out four times and then RCF started to send the traffic. A line of 4646's separated the repetitions.

The above was sent on April 29, 1992, so it would seem to be a valid conclusion that the QSF designators were for the May Day holiday period plus the next two Sundays. Just how many schedules QSF 3 and WSF 2 refer to has not been determined.

It was interesting to see a similar TIKAS message show up on the large network but in a slightly modified version. This particular transmission was to SPK.

1/5, 2/5, 9/5 – QSF4
3/5, 10/5 – QSF3

Here is an example of the callup by RCF, Moscow MFA. The callup sentence is sent for five minutes followed by the traffic.

4646464646464646464646464646464646464646464646464646464646
CQ CQ CQ DE RCF CQ CQ CQ DE RECF
FOR JUA RXX FQX UDZ21 CAZ RPO FRU BLA RCX81 FJN RKG RKM BXL
URO BNV EWZ42 NR 08 ALL GR 112

4646464646464646464646464646464646464646464646464646464646464646
CQ CQ DE RCF CQ CQ CQ DE RCF
FOR JUA RXX FQX UDZ21 CAZ RPO FRU BLA RCX81 FJN RKG RKM BXL URO BNV EQZ42 NR 08 ALL GR 112
4646464646464646464646464646464646464646464646464646464646464646
4646464646464646464646464646464646464646464646464646464646464646

NR 08 GR 112
NR 08 GR 112
MCETS IDPLB VHTHR JHZFJ UTEBG HPYCJ OOSER EDKCO JMAJE LWTNG
HZBOF LCAOW OVTGG UUTEO MNLWY ZEDMG OLXLB PRHAE TVESW
DIRXE GWTQT FVYCV JRWYW OHGOI WIHED HJVAF FTHCH YSWPB
GNHPG FXVHT GDKCU UOLRW FNFXU XTZOR ALBWK QHPGO
LSZAX CJEDK JIKEK DKFHV BTDMR WPWKL RGIAT FCMTC MNUEB
EVGEO YYYYH HRFYJ YCGLE CNKLF QNCAF JWDQ KDQUW
XNIHN TIJGE ZMSMW SQMYD RTWIM VPTII SOODF VLTGS NYARY
RZUSY RPYHN DRSCC IJTOY XCYDH PATIF BIHRZ QEAYG JPVKB
EYXLZ RUHQW LTEQK FLXQQ NNOQR NUOZN XOPPR YVMJS
GEKLO XTEXW QKOOF TKTGH CNWWX XHJLF IFEVJ CFBZS TCFLY
WMWDI PTXZN CYQFC RIJNB ORTWW LYBYU BPJOS PSZNO
QFIKH RUJAE ADVXE WEBZV CZHPG NFMMN PBGQT ZQUSR
1PLHZ CMGRT WZVQ EZNAX KQCLJ ICHMV

QRU SK QRU SK

Here is a sample callup indicating traffic:

VKX VKX VKX QTC VKX VQK VKX QTC VKX VKX VKX QTC VKX VKX VKX QTC
46464646464646464646464646464646464646464646464646464646464646464

(The above sequence is repeated approximately 10 times).

11199 00166 00000 12241 00029
14300
ALL QTC 1 ALL QTC 1
QRU QRU SK SK

To VKX

		Heading Groups						Text
Day	Date	1st	2nd	3rd	4th	5th	Tfc	Last Grp
Wed	02 Dec 92	11199	00166	00000	23480	00029	QSL	21800
Tue	15 Dec 92	11199	00166	00000	15352	00029	QSL	21900
		11177	00166	00000	15353	01941	5F	11111
Wed	20 Jan 93	11199	00166	00000	20013	00029	5F	18890
Sat	06 Feb 93	11199	00166	00000	06024	00029	QSL	01600
Mon	15 Feb 93	11199	00166	00000	15030	00029	QSL	02000
Fri	05 Mar 93	11177	00166	87813	05046	00749	5L	OROWI
Tue	09 Mar 93	11177	00166	00000	09047	00031	5F	11111
		11199	00166	00000	09048	00029	QSL	03100
Wed	24 Mar 93	11199	00166	00000	24067	00029	QSL	04400
Fri	26 Mar 93	11177	00166	87967	26070	01009	5L	UEOPW
		11177	00166	00000	25068	12211	5F	11111
		11199	00166	00000	26069	00029	QSL	. . . . .
Wed	31 Mar 93	11199	00166	00000	30077	00029	QSL	04800
Wed	07 Apr 93	11199	00166	00000	07078	00029	QSL	04900
		11177	00166	99184	07079	02259	5L	O.UUU
Wed	21 Apr 93	11177	00166	58281	21094	01139	5L	UIIIO
		11166	00166	00000	20091	02901	5F	11111
		11199	00166	00000	21092	00029	QSL	05700
		11177	00166	00000	21093	01061	5F	11111
Wed	21 Apr 93	11177	00166	00000	21093	01061	5F	11111
Fri	21 May 93	11177	00166	00000	21122	02281	5F	11111
Mon	24 May 93	11199	00166	00000	23124	00029	QSL	07300
Tue	25 May 93	QRU						
Sat	29 May 93	QRU						
Fri	04 Jun 93	11199	00166	00000	03130	00029	QSL	076 . 0
Wed	09 Jun 93	11199	00166	00000	08138	00029	QSL	08100
Tue	22 Jun 93	QRU						
Mon	28 Jun 93	11199	00166	00000	27156	00029	QSL	09100
Tue	06 Jul 93	11177	00166	61800	06161	02439	5L	OEITO
Mon	19 Jul 93	11199	00166	00000	18174	00029	QSL	10100

		11199	00166	00000	19175	00029	QSL	10200
Tue	27 Jul 93	QRU						
Wed	18 Jul 93	QRU						
Wed	04 Aug 93	11177	00166	11100	04187	04031	5F	. . . . .
Fri	06 Aug 93	11177	00166	46928	06192	00539	5L	OEORO
		11199	00166	00000	06191	00029	QSL	11400
Sat	07 Aug 93	11199	00166	00000	06193	00029	QSL	11500
		11199	00166	00000	07194	00039	QSL	11600
Wed	18 Aug 93	11177	00166	15021	18206	02639	5L	IAUEO
Tue	24 Aug 93	QRU						

		Heading Groups						*Text*
Day	*Date*	*1st*	*2nd*	*3rd*	*4th*	*5th*	*Tfc*	*Last Grp*
Thu	16 Sep 93	QRU						
Wed	22 Sep 93	QRU						
Fri	24 Sep 93	11199	00166	00000	24225	00029	QSL	13400
Mon	27 Sep 93	11199	00166	00000	26229	00029	QSL	13600
Tue	28 Sep 93	111 . .	00166	89799	28230	00669	5L	UAIEZ
Tue	05 Oct 93	11199	00166	00000	04236	00029	. .	. . . . .
Wed	06 Oct 93	11199	00166	00000	06237	00029	QSL	14100
Mon	11 Oct 93	QRU						
Wed	13 Oct 93	11199	00166	00000	12241	00029	QSL	14300
Wed	20 Oct 93	QRU						
Fri	29 Oct 93	QRU						
Mon	01 Nov 93	QRU						
Tue	02 Nov 93	QRU						
Sat	06 Nov 93	QRU						
Tue	09 Nov 93	QRU						
Thu	10 Nov 93	QRU						
Mon	15 Nov 93	11199	00166	00000	14258	00029	QSL	16100
Wed	24 Nov 93	QRU						
Fri	10 Dec 93	QRU						
Sat	11 Dec 93	QRU						

To MISCELLANEOUS

Day	Date	Heading Groups 1st	2nd	3rd	4th	5th	Tfc	Text Last Grp
	25 Dec 92	11177	80061	00000	25258	07241	5L	. . . . .
		11177	80061	00000	25395	07891	5L	. . . . .
		11177	80061	00000	253 .6	07971	5L	. . . . .

Note: The first message was on 16153 kHz at 1355 UTC. The second message was on 5072 kHz at 1800 UTC. The third message was on 5076 kHz at 1815 UTC.

Day	Date	1st	2nd	3rd	4th	5th	Tfc	Text Last Grp
	Sep 92	11177	10190	00000	14835	02361	. .	. . . . .

Note: This message was on 16215 kHz. No time given.

Russian Cipher Systems

In his book, ***Tower of Secrets***, Victor Sheymov gives some information about various Russian cipher systems. Here is a quick summary:

Regular	Used at low traffic volume sites.
Station Chief	His personal cipher.
Emergency	Used when other ciphers are destroyed or compromised.
Temporary	Used at newly opened embassy when secure *referentura* is not yet available.
Miniature	Used by communications officer assigned to delegations such as Olympic teams.
Superminiature	Used by "illegals" and other field agents.

The model of crypto machine used by KGB, GRU, and MFA communications to and from foreign posts was called the "apatit." The "referentura" mentioned above is a vaulted area with walls of reinforced concrete. Inside are reading rooms and soundproof rooms for conducting conversations of a sensitive nature. An innermost portion has separate rooms for the cipher clerks and cipher equipment for each component (KGB, GRU, and embassy). There is an also an inside vault which the houses the safes of the three component.

According to explanations given by Sheymov, in a manual cipher system a plaintext message was encoded through the use of a code book. That version was next enciphered by using a gamma (one-time pad). In a machine system, the plaintext was entered into the machine via the keyboard and the resulting enciphered version combined with a tape of random characters to form a super-enciphered text for transmission.

Mr. Sheymov's book reveals many very interesting details of some phases of Russian communications. The author defected to the United States in 1980, and for ten years the Soviets did not realize that he and his family were alive. Before his defection, he was in charge of Soviet worldwide cipher communications security.

Soviet Legal Holidays

1 January	New Year's Day
8 March	International Women's Day
1-2 May	Day of International Worker's Solidarity
9 May	Victory Holiday
7-8 November	October Holiday#
5 December	USSR Constitution Day

It is not known if all of these holidays will be observed by the Russian government. In the past, in addition to the above, a reduced schedule has been noted during the Christmas season.

 # The revolution actually took place in October, but with the change to the Gregorian calendar from the Julian after the Russian revolution, this particular holiday is celebrated in November.

Cuban Diplomatic Traffic

About 10 years ago, I received some information from a person who wished to remain unidentified. He described a cipher system in use by the Cuban Ministry of Foreign Affairs (MINREX) which was used for enciphering some elements of message headings. The individual also mentioned there was a CW/RTTY link suspected of handling traffic between the Ministry of Defense in Cuba and the Cuban troops fighting in Angola at that time.

When I first monitored this latter link, it was immediately apparent that the same procedure of enciphering part of the message headings was being followed and that the cipher system utilized for this was the same one used for the headings of the Cuban diplomatic messages.

Here are the cipher and plain alphabets for the monoalphabetic system:

CIPHER: A B C D E F G H I J K L M N O P Q R S T U V W
X Y Z

PLAIN: Z I A M Y D O Q ? V R S J B G P U T C F H L W X N E

This practice continued through the years for the Cuban diplomatic traffic and was still in use for that traffic when this book was prepared in 1994. It remains a mystery to me why it is considered necessary to encipher certain heading details since the data is also duplicated in clear text in the heading.

For comparison purposes, let's take a look first at a heading from the Angola/Cuba military traffic and then one from a much more recent Cuba diplomatic message heading.

CIPHER: JBYRBFGLZDBVZSQCRKGSBYRGSBYSGCZKPRZCZFVZMCOQCKZ

PLAIN: VINTIDOS MIL CUATROCINTOCINCOA RPT A DL JAGUAR

CIPHER: CZSGNCZZZZZZZZZZ (and into another type of encryption)
PLAIN: A COBRA

TRANSLATION: NR22405A REPEAT A FROM JAGUAR TO COBRA

Apparently because the cipher Z is also used as a separator, the letter E is omitted from the plain text words in the headings.

Now we will look at a Cuba diplomatic message. This message was copied on December 9, 1992:

HABANA/ 8/12/92 MH
RELACION NICARAGUA
1658/ ?LGMI 101 ORD
MINREX 202

TRASMITE MH
HORA 1925 RECIBE ROP

ZZZZZZZZZZZZ
DNCSQNCZYBSCKCOQCZGKFS
KGZZIVODBZZZZZZZZZZZZZZZZZ (and into the encrypted mode)

Here is the breakout of the simple cipher portion of the heading:

CIPHER: DNCSQNCZYBSCKCOQCZGKFS
PLAIN: MBACUBA NICARAGUA ORDC

CIPHER: KGZZIVODBZZZZZZZZZZ etc.
PLAIN: RO ?LGMI

Again, you will observe that plain text E is omitted from the deciphered words in the heading.

The ?LGMI group is suspected of being a synchronization feature which ensures that sending and receiving crypto modes for the message text are in phase with each other. Each cipher message has such a group in the plain text heading, and it is usually repeated in the enciphered portion of the heading.

It is not known what the plain text CRO stands for in the above decrypted portion (cipher characters SKG). The ORD which precedes the CRO is probably the message precedence which in this case is ORDINARIO. This precedence is possibly equivalent to the U.S. precedence of ROUTINE.

I do not copy as much Cuban diplomatic traffic as I used to, but here are addresses I noted in traffic monitored during 1993.

CONSULCUBA HO CHI MINH	
EMBACUBA AFGANISTAN	
EMBACUBA ANGOLA	CLP45
EMBACUBA ARGELILA	
EMBACUBA BENIN	CLP15
EMBACUBA BURKINA FASO	
EMBACUBA CAMBODIA	
EMBACUBA CONGO	CLP7

EMBACUBA COREA	
EMBACUBA EGIPTO	
EMBACUBA GHANA	
EMBACUBA GUINEA BISSAU	CLP4
EMBACUBA GUINEA CONAKRY	CLP8
EMBACUBA GUYANA	CLP55
EMBACUBA IRAQ	CLP67
EMBACUBA LAOS	
EMBACUBA LIBANO	CLP3
EMBACUBA MOZAMBIQUE	CLP25
EMBACUBA NICARAGUA	CLP65
EMBACUBA NIGERIA	CLP23
EMBACUBA PANAMA	CLP2
EMBACUBA PERU	
EMBACUBA SIRIA	CLP6
EMBACUBA TANZANIA	CLP18
EMBACUBA UGANDA	
EMBACUBA VIET NAM	CLP22
EMBACUBA YEMEN	CLP9
EMBACUBA ZAMBIA	
EMBACUBA ZIMBABWE	

It was not possible to match up all of the above addressees with their appropriate CLP callsigns because many of the messages copied were of the "CIRCULAR" type (multi-addressees) and were sent to one station for relay to certain other stations.

A Suspected Cuba/Russian Embassy in Washington Link

One of the stations in the two-way link discussed previously is suspected of being in Cuba at the Russian transmit/intercept facility and the other station is located in the Russian Embassy in Washington, DC.

Monday-Saturday schedules have been observed at 1415 and 2115 UTC. During heavy traffic periods, they will often hold a schedule at 1815. Cuba usually comes on frequency about 15 minutes prior to the schedule. The initial callup by Cuba is "WFO WFO WFO" and most times is in FSK Morse. Cuba will indicate QTC usually followed by "NW NW" and he shifts to RTTY 75/500 and sends some "64's." Apparently, the Russian embassy then tells Cuba to go ahead and Cuba sends the traffic. As an additional identification feature, the second group of the message heading on traffic destined to "WFO" will always be 00125.

The Russian embassy in Washington has only been heard one time and this was on May 20, 1993 on 13379.9 kHz at the 2115 UTC schedule. Since that time, new operating frequencies have been placed in use effective June 1, 1993. A "TIKAS" operator note was copied and it listed the following the following times/frequencies:

TO WFO	1200-2300	12455/14726/13448
	2300-1200	12455/10233/8109
FROM WFO	1200-2300	10856/13372/15559
	2300-1200	10856/9369/7735

Although the operator service message indicated 14726 kHz, he is on 14736 kHz. This could have been a garble, but there is QRM in the vicinity of 14726 kHz so perhaps he moved up some time ago. This signal is the primary and has been in use ever since I first heard this particular schedule. In addition to the frequencies given above, Cuba has also utilized 10231.4, 10842, 11404, 14014, 16232, 16446, 16888, 16896, and 18713 kHz.

I suspect the Moscow/Cuba communications are via satellite. In 1992, I monitored HF RTTY communications one time which appeared to be Moscow/Cuba. This was because my West Virginia location is probably beyond ground wave reception, and the sky-wave signal from Washington should be "skipping" overhead and ought to be inaudible. The Russian embassy in Washington is very likely utilizing a highly directional antenna array, so any signal would be off the "back end" of their antenna and thus very weak.

To copy this RTTY traffic on my M-7000 terminal, I found it necessary the turn the UOS function *off*. And the tuning sequence is definitely "64's," not "RY's."

3

Four Puzzles: SLHFMs, FEMA, the KKN Stations, and the NCS

WHILE NUMBERS STATIONS have long been the best-known (and perhaps most "glamorous") part of underground radio, they are by no means the only unusual things you can hear on shortwave radio. One persistent mystery has been the numerous stations that do nothing more than continuously repeat a single letter of the Morse code. Two communications networks we will examine in this chapter—the Federal Emergency Management Agency and the U.S. Department of State's "KKN" radio system—are on the surface used for routine, non-covert government communications. However, both have some interesting aspects that deserve further scrutiny. There is another government radio system, the National Communications System (NCS), that may be intimately involved with many of the mysteries discussed in this book.

Some Initial Observations

Before discussing the communications of single-letter high frequency markers (SLHFMs), the Federal Emergency Management Agency (FEMA), and the U.S. Department of State's KKN stations, I believe it is appropriate to pass along some observations.

One relatively recent innovation in the communications field is the capability of a station to transmit from different locations throughout the world. I recall hearing Andrews AFB working Air Force One several year s ago, and Andrews transmitted through a number of transmitters, one after the other, which were located at different bases including some overseas bases. Apparently, the site initiating the contact routes its transmissions via satellite to another location where the signal is patched through to an HF transmitter and is then sent out via that emitter. Of course, the necessary satellite equipment would have to be available at such site. This procedure might explain why certain transmissions, such as those mentioned in the opening paragraph, seem to be coming from sites other than the one generally regarded as the station location.

On another point, I have noticed an apparent misunderstanding by a few hobby publications concerning the various names used by the U.S. Army COMINT organization. For clarification, here is an outline of the sequence of the various titles.

In 1929 the Signal Intelligence Service was created and the name lasted into 1942. From June, 1942 to July, 1942, there were three different, unidentified titles assigned and each was then promptly replaced. From July, 1942 to July, 1943, the organization was known as the Signal Security Service. Then in July, 1943, the name was changed to the Signal Security Agency, and it remained so until the end of World War II. On September 15, 1945, the Signal Security Agency plus the army field cryptanalysis units and the Signal Corps cryptography functions were all combined to form the Army Security Agency (ASA), and the latter title has continued as the official name.

Single-Letter High Frequency Markers

Single-letter high frequency markers (SLHFMs) have been a persistent mystery since the early 1960s. These generally do nothing more than continuously repeat a single letter of the alphabet in Morse code. However, some are interrupted now and then for five-digit groups or data bursts. Most use CW, although a few have used frequency-shift keying. Some of the Morse code characters used have been from the Cyrillic alphabet. The locations of the various SLHFMs have long been subjects of debate. Most of the evidence points to locations within the former USSR, particularly the Asiatic regions.

Let's briefly run through the history of these unusual transmissions. Initially, they were labeled "SLHFBs," with the "B" standing for "beacon." However, I and other monitors believed that was not the best definitive title for these strange broadcasts. Thus, I started calling them SLHFMs in my ***Popular Communications*** column and elsewhere.

In the early 1970s, a few utility monitors pooled their efforts to see what could be deduced regarding these mysterious transmissions. As data was gathered, some definite facts came to light. Rough DF bearings by use monitors showed Soviet territory as being the location of some of the SLHFMs. In the mid-1970s, it was determined that Soviet personnel were constructing facilities at the Cienguegos, Cuba naval base capable of servicing Soviet ballistic missile carrying submarines. The United States protested to the Soviet Union that this constituted a violation of the 1962 U.S./Soviet understanding regarding the Cuban missile crisis, which denied the USSR the use of Cuba for installation of offensive weapons systems on the island. This construction was supposedly abandoned in the later 1970s.

A "W" station had surfaced in the mid-1970s on 3584 kHz causing QRM in the 80-meter ham band and FCC direction-finding placed the transmitter in Cuba. A formal protest was made to the Cuban government. It is not known if what followed was a result of the protest, but the signal went off the air for several months, but was then again active, this time on 3533 kHz. Sometime in 1978, the signal again left the air. With the similarity in time frames of the above mentioned events, it could be speculated that there was a connection between the Soviet naval activity in Cuba and the "W" signals.

The next observance of the "W" signal was on 8002 kHz. It is not known if there were additional DF fixes at that time which confirmed the "W" transmissions were still coming from Cuba. This appearance was most interesting because the "W" marker was now being interrupted by transmissions like "OT7F DE YAAX," followed by a resumption of the "W" marker, only to be broken yet again later by "DE YAAZ OK SK SK." From this it would certainly seem evident that the four-character callsigns were definitely related to the "W" activity somehow .

Monitors reported that SLHFMs often seem to be operating in some sort of synchronization with each other. It was not unusual to find two or more SLHFMs within a kHz or two of each other. Other "groups" of SLHFMs could be found operating in several different frequency bands at the same time. Monitors also noted variations in the speed at which the letters were transmitted and the spacing between letters, as well as slight variations in the transmitter frequency over time. (Not to be overlooked is the possibility that such variations were caused by sustained changes in the power system voltage levels.)

As we entered the 1980s, a marker was observed with the Russian Cyrillic letter IO, which in Morse code sounds like IM (dot-dot-dash-dash). There were now reports that more and more monitors were seeing the signal ceasing and short messages of five-figure groups transmitted, then repeated several times, followed by a return to the single-letter designator repetition. In 1982, an unknown individual obtained the results of some U.S. military direction-finding of the "K" marker which operated on 9043 kHz. The fix placed the site in the USSR near Khabarovsk.

A summary of the findings up to early 1983 appeared as a section of the ***SPEEDX Reference Guide to the Utilities***. The next published compilations of SLHFM data were prepared by William Orr and appeared in ***Popular Communications*** issues of December, 1984, January and February 1985, and June, 1986.

Another development occurred in 1986 when the FCC released some fixes from its DF operations showing the general location of some of the SLHFMs as follows:

Marker	*Location*
C	Moskva
D	Odessa
O	Moskva
P	Kaliningrad

S	Arkhangelsk
U	Area between Murmansk & Amderma
Z	Mukachevo

There are those who believe not all SLHFMs operate (or operated) from Soviet territory. There is, of course, the "W" signal previously mentioned as being located in Cuba. Other SLHFMs have been noted operating at times and on frequencies unlikely to support propagation from the USSR. For example, the late Dave White, from his location in Maine, reported he could hear a "K" SLHFM on 8790.5 kHz day and night with a S9 signal, meaning it was highly unlikely that the signal was coming from the USSR all the time.

One possibility is that some of these signals could be originating from different Soviet naval bases around the world, or perhaps even from seagoing Soviet ships. This latter possibility should certainly not be discounted.

What are we to think of this part of the puzzle? Frequently, the short messages have been preceded by several repetitions of the callsign "UMS" which while supposedly the Soviet Navy station at Moscow, was actually believed to be a general Soviet naval callsign used by transmitter sites at Petropavlovsk and Murmansk. Also, monitors were seeing appearances of other Soviet naval callsigns interrupting the marker transmissions. "ROD" and "RMP" were monitored along with other Soviet naval callsigns. I think it is also noteworthy that each of the identified locations of SLHFMs had a Soviet naval unit of some type in the particular area.

Through propagation studies plus additional DF readings, most identified locations tended to be in line with previous determinations, with the exception of Kharbarovsk which was subsequently refined as probably being further north, near Petropavolovsk.

Much speculation has been devoted to the purposes of SLHFMs, with most centering on their use by the Soviet/Russian Navy as navigational markers, or perhaps for signaling when a vessel is out of range of a communication satellite. Other suggested possible uses were as radio propagation beacons, weather data transmitters (with the weather encoded in five-digit groups), or as standby transmitters in the event of loss of satellite communications systems. Another possibility suggested was that SLHFMs were used to transmit a type of measurement such as water levels in rivers, lakes, and reservoirs in the former USSR.

More and more loggings were being reported of the four-character callsign transmissions which operated in much the same fashion as the SLHFMs. The callsign would be sent over and over for extended periods and then be interrupted for traffic. Some of these messages involved Cyrillic characters while others had texts of five-digit groups, five-letter groups, four-letter groups (full numbers), and four-letter groups (cut numbers). Many direction-finding "shots" were made from California and Australia, resulting in the belief that one of these stations, YQBF, was located in, or near, Vladivostok. Here again, the location hosted a large Soviet naval base.

Another factor noted by monitors was that some of the traffic showed the Moscow time zone in the date/time groups while other messages showed the time zone, in the date/time groups, which included portions of Czechoslovakia and East Germany. Thus it would clearly appear there was Bloc affiliation with these mysterious communications. Another significant point, discovered when looking through old traffic files, was the "O" signal made a schedule change of one hour on September 28, 1987 when the USSR changed the clock back from summer time.

For recognition purposes, here are examples of several types of messages passed by these four-character callsigns:

DE VIZ6 QTC 7 20 26 0205 BT 067 SFCJ BT
YAJUL IQXOR EJKCA 38192 DAENV XBIDZ 98102 WYJKA YQIMX
DZHGR FJCOA WKNOX XTQFA VWRKA IRGNC 21760 94137 LGDAC
SBXNW FQUCE BT VIZ6 AR

This intercept consisted of several short messages in cut-numbers groups:

MKLI MKLI BT 7ARA 4DD3 TYTD NTNY 3734 TTAT NTN5 BT
DRQJ DRQJ BT 66CA TUT4 UTT6 U33N 7456 D677 5363 BT
YKPU YKPU BT D77A 3534 564U D4D4 D7A7 56D7 4AUA BT

Four more similar messages followed and then the format changed somewhat, and you will note instances in the heading of this message where a group was sent as full numbers and then repeated as cut numbers:

PP 8528 D5UD PP NR 38 NR 38 GR 18 GR 18 21/2000 UA/UTTT
NOAU NOAU TO 9115 NAA5 BT
6588 1158 8850 2158 6634 1794 7527 9165 6411 6017
6926 3445 1116 6926 2632 9206 5906 5604 BT AR TKS
QRU K

The cut number system used in the foregoing messages was AU34567DNT = 1 to 0.

I guess it was 1990/1991 that I received an odd letter (unsigned, of course) in which it was stated that some of the traffic sent by SLHFMs and the four-character callsigns was what was called "direction-finding tip-off" activity. No explanation was given, but one would have to assume what he meant was that the enciphered messages gave frequencies, operating times, station characteristics, etc., of various communications targets for which it was desired that the recipients of the traffic (DF sites) would perform direction-finding operations to locate the particular transmitters.

Toward the end of 1989, some SLHFMs were no longer being heard, but there were many that were still active. Likewise, there was continued activity by the four-character callsigns. The collapse of the USSR did not put an end to SLHFMs, although there were periods of low activity in the early 1990s. Both SLHFMs and four-character callsigns are still being reported by monitors at the time this book was being prepared in 1994, and they both remain one of the biggest mysteries found on shortwave.

Federal Emergency Management Agency (FEMA)

The Federal Emergency Management Agency is an outgrowth of the old "civil defense" programs of the 1950s designed to prepare for the possibility of nuclear war. FEMA's scope of responsibility today includes the following:

- Supporting state and local governments in a wide range of disaster planning, preparedness, mitigation, response, and recovery efforts.
- Coordinating federal aid for presidentially declared disasters and emergencies.
- Coordinating civil emergency preparedness for peacetime radiological accidents, including those at nuclear power plants, and hazardous materials incidents.
- Providing training, education, and exercises to enhance the professional development of federal, state, and local emergency managers.

FEMA offers on-campus and off-campus training programs through its National Emergency Training Center located in Emmitsburg, MD. Located at the Center's campus are the Emergency Management Institute, which offers training in the areas of national security, civil defense, technological hazards, natural hazards, professional development, and the exercise process, and the National Fire Academy, which offers training to advance the professional development of fire service personnel and of other persons engaged in fire prevention and control activities.

While FEMA today gets most attention from its efforts to deal with natural disasters such as hurricanes and earthquakes, there are still links to its civil defense past. According to an Associated Press report, FEMA spent most of its money in the 1980s on a top secret program to enable the government to survive a nuclear attack. The program, which was built around a vast communications network, included a fleet of 300 vehicles in five mobile units scattered from the state of Washington to Massachusetts and from Colorado to Georgia. The communications project supposedly began during the Carter administration and was expanded during the Reagan years.

FEMA can be heard using both voice (typically USB) and CW modes. Routine tests of FEMA's voice networks are often heard on frequencies such as 10493 kHz. FEMA coded transmissions in CW have been frequently logged by monitors on frequencies like 3378 kHz. Currently, the traffic is composed of groups of three-alphanumeric characters with the messages repeated numerous times. All known currently active FEMA frequencies are included in the Underground Frequency List. While frequencies have been listed as USB or CW frequencies, either mode may be heard on a frequency. In some cases a description of the communications typically heard on a frequency is given. Note that some FEMA stations have been heard in contact with stations from other agencies of the U.S. government such as the Department of Energy.

FEMA's operations are organized into regional districts, as shown in the accompanying map below. The stations at the district offices are assigned calls from the WGY900 to WGY915 sequence. Stations with callsigns in the WGY920 to WGY998 sequence belong to state and local agencies affiliated with FEMA. Stations assigned to each district can be identified by the last number in the callsign; for example, the call WGY954 would belong to the fourth district.

The FEMA network control station is WGY903. Alternate control stations include WGY904, WGY905, WGY906, and WGY908. Station WGY915 in Arlington, VA, is the FEMA outlet associated with the National Communications System, which will be discussed later in this chapter.

FEMA Regional Offices are:

Region I (Boston)
442 J.W. McCormack, POCH
Boston, MA 02109
(617) 223-9540

Region II (New York)
26 Federal Plaza
New York, NY 10278
(212) 264-8980

Region III (Philadelphia)
Liberty Square Bldg. (Second Floor)
105 South Seventh Street
Philadelphia, PA 19106
(215) 597-9416

Region IV (Atlanta)
1371 Peachtree Street, N.E.
Suite 700
Atlanta, GA 30309
(404) 347-2400

Region V (Chicago)
175 W. Jackson Blvd., Fourth Floor
Chicago, IL 60604
(312) 431-5501

Region VI (Dallas)
Federal Regional Center, Rm. 206
Denton, TX 76201
(817) 898-9104

Region VII (Kansas City)
911 Walnut Street, Rm. 300
Kansas City, MO 64106
(816) 374-5912

Region VIII (Denver)
Federal Regional Center, Bldg. 710
Box 25267
Denver, CO 80225-0267
(303) 235-4811

Region IX (San Francisco)
Building 105
Presidio of San Francisco
San Francisco, CA 94129
(415) 923-7100

Region X (Seattle)
Federal Regional Center
130 228th Street, S.W.
Bothell, WA 98021-9796
(206) 487-4604

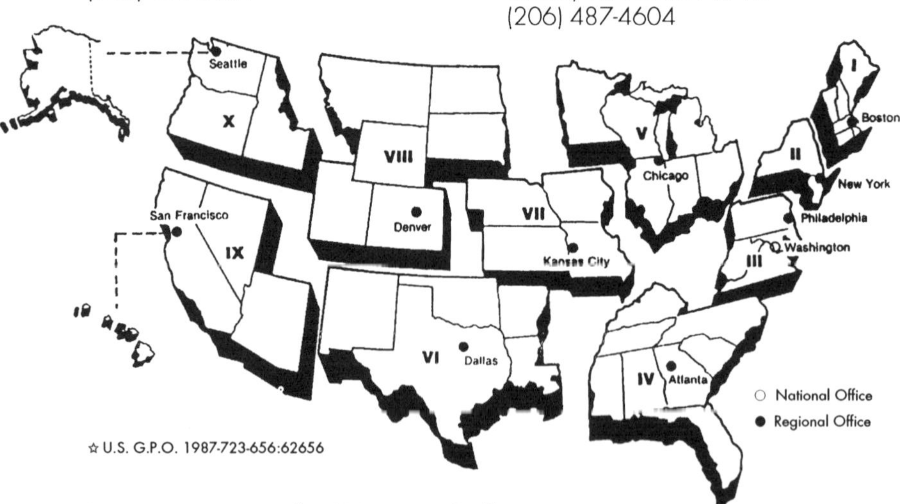

Figure 3-1: Here is a map of FEMA regional offices, as reproduced from an official FEMA publication.

The callsign of the special FEMA facility at Mt. Weather in Virginia is WGY912. Mt. Weather is the principal relocation facility for government officials in the event of nuclear war or the threat of imminent attack. The WGY912 callsign has been seen on many transmissions but several monitors who have listened to these broadcasts believe that some of the signals were actually coming from other locations. This belief was based on propagation factors and rough DF fixes. For example, the late Dave White of Maine reported in 1990 about monitoring WGY912 on 10870 kHz with three-character traffic. Dave used an indoor directional loop antenna. While such an antenna is not sufficiently accurate to permit exact determination of the location of a station, it can give a good idea of the general area in which a station is located. Dave reported that WGY912 apparently switched to a new transmitter site each 15 minutes. While all loop bearings were in the 230° to 280° range from his location, there were sudden and unambiguous changes in bearing direction on a regular basis at 15 minute intervals. Dave believed that at least four different sites were in use. Another clue the WGY912 is used by transmitter sites other than Mt. Weather was supplied by Harry Helms. Several times in 1992, Harry heard "WGY912" on 3378 kHz as late as 1400 UTC at his home in southern California, a time highly unlikely to support propagation from Virginia on that frequency.

The FEMA channels are worth monitoring whenever a natural disaster has taken place or is anticipated (such as a hurricane approaching a coastal area). Moreover, it is clear there is more to FEMA's communications facilities than first meets the eye. In particular, there has been much speculation as to the purpose of the coded CW groups transmitted on some FEMA frequencies. Some SWLs have wondered if perhaps there was some link between FEMA and other radio mysteries, like numbers stations. There is no hard evidence supporting that speculation, but FEMA is certainly fascinating in its own right.

The "KKN" Stations

Many SWLs around the world are familiar with the so-called "KKN" network of stations. This network gets its name from the "net control" stations, KKN39 and KKN50, which are listed as belonging to the U.S. Department of State in Washington, DC. Other well-known stations in this network include KRH50 (U.S. embassy, London) and KWS78 (U.S. embassy, Athens, Greece). KKN50 can be easily heard on 6925.5 kHz most evenings repeating a CW marker that goes something like "QRA QRA QRA DE KKN50 KKN509 KKN50 QSX 6/13/17" for hours on end. Sometimes around the hour (and, more rarely, near the half hour) these marker signals will be replaced by hand-sent CW of either the marker or brief coded groups, and the marker tape loop will then resume. SWLs have even managed to secure QSL cards for reception of these markers. Figure 3-2 shows a prepared QSL card returned to Patrick O'Connor of New Hampshire indicating verification of his reception of KKN44, a station that was listed as operating from the U.S. embassy at Monrovia, Liberia.

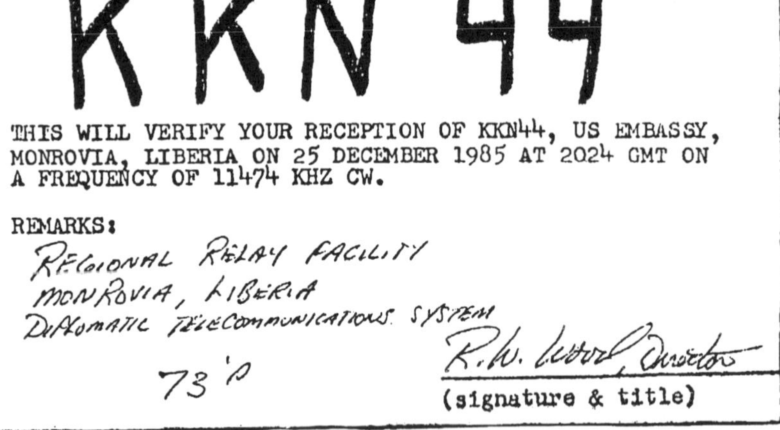

KKN 44

THIS WILL VERIFY YOUR RECEPTION OF KKN44, US EMBASSY, MONROVIA, LIBERIA ON 25 DECEMBER 1985 AT 2024 GMT ON A FREQUENCY OF 11474 KHZ CW.

REMARKS:
REGIONAL RELAY FACILITY
MONROVIA, LIBERIA
DIPLOMATIC TELECOMMUNICATIONS SYSTEM
73'S

R.W. Wood, Director
(signature & title)

Figure 3-2: KKN44 signed and returned this QSL to Patrick O'Connor of New Hampshire; Patrick prepared this card for the station's use. Note that the operator signing the card described the station as a "regional relay facility."

For years, these stations were said to be used for low-level routine diplomatic traffic that did not need to go through faster or more secure channels. However, monitors soon noted that none of the KKN stations seemed to engage in any sort of two-way communications with each other. Instead, they simply transmitted the CW marker with sporadic interruptions for hand-sent code. This generated rumors that the KKN stations were actually used for some other purposes, like espionage. Even without resorting to speculation, the continued existence of the KKN stations seemed a bit odd in an age of secure, faster communications via cable and radioteletype. Did the U.S. Department of State really need to communicate with its London embassy in CW via shortwave radio? It was if the KKN network represented start-of-the-art communications circa 1935.

I guess it was sometime in 1991 that I received a letter signed "The Old Communications Man" which outlined certain details concerning the KKN stations. As I remember his comments, the U.S. Department of State had a communications network linking it with the various American embassies and consulates around the world. He said this network was called the Diplomatic Telecommunications System (DTS). At each site, the DTS passed traffic for the embassy, consular service, Agency for International Development (AID) personnel, U.S. defense attaches, U. S. Information Agency (USIA) personnel, and others. He did not specify who the "others" were!

My correspondent claimed that some sites had satellite links but that most sites had HF high speed crypto links and that the KKN marker frequencies were nothing more than operator service channels where normally the operator exchanges consisted merely of QSY instructions, QSA reports, etc. He said these frequencies were kept open through the use of QRA/QSX makers, and he stressed that only in extremely rare cases would any actual message traffic ever be passed on these marker frequencies. He added there was an after hours alerting method similar to the aeronautical Selcal procedure.

Another point I recall was his mention of KKN sometimes sharing transmission facilities. Here again, he did not elaborate. He indicated that by using equipment of higher power at alternate locations, it was much more likely distant embassy stations would be able to hear after hours transmissions if this were required. This is an important point that should be kept in mind, as will be clear later in our discussion.

I must point out that several items in the letter I received coincided with some of the information Larry Van Horn of *Monitoring Times* had apparently received just about the same time. His source reportedly was in North Carolina, while the envelope I received had a northern Virginia postmark on it. I wish I had kept the letter and envelope so that Larry and I could have compared the handwriting. I might add that this would not be the first time that an individual has provided similar details to both Larry and me.

While there is no way to confirm the claims of "The Old Communications Man," his statements do agree with much of the evidence that has accumulated over the years regarding the KKN stations. There are several intriguing things about their operation.

One such thing is how significant technical problems with a transmitter can continue for some time before being corrected. In November, 1987, Harry Helms noted station KKN39 on 4957.2 kHz with a very rough, raspy CW signal. The keying had a strong AC component present. Rather than being quickly fixed, the problem continued for five more days. Just before the problem was repaired, KKN39's signal had drifted to 4956.7 kHz and the keying sounded like a "buzz." There were other problems with KKN39 during this period. In the February, 1988 issue of *Monitoring Times*, Ken Cohen of Maryland reported hearing KKN39 on 14001.8 kHz (in the 20 meter ham radio band!) calling station KUO029. Ken described the signal as having "a chirp and a hum." While no date was given for Ken's reception, the lead time for publication of *Monitoring Times* indicates it took probably place about the same time as Harry's. These problems—the tone of the CW signal and off-frequency operation—should have been quickly noticed by any operators monitoring KKN39's signals.

Harry Helms began intensively monitoring the KKN stations after the reception noted above. He was able to observe that some of the stations observed consistent operating schedules on various frequencies. For example, during October, 1988 Harry reported that KKN50 on 6925.5 kHz would begin sending its QRA marker at 0100 UTC daily and would sign off at 0600. There were some minor variation of a few minutes day to day, but operations were otherwise very consistent. He noted that the only interruptions came around the hour and half-hour. In some cases, the "canned" CW marker was sent by hand (apparently using an electronic keyer) at speeds of over 25 words per minute. At other times, short random groups such as "4ISDY00" would be sent. And at other times KKN50 would call other stations with callsigns such as KWY54 and KPH7. All interruptions to the QRA markers were brief, however, and the marker would usually resume after an interruption of less than a minute.

Harry next resolved to determine the sign off time of KKN44, listed as being the station at the U.S. embassy in Monrovia, Liberia. This station put a good signal each evening on 4886 and 7652 kHz into Harry's southern California listening post. The 4886 kHz frequency left the air at 0600 each evening, Harry felt 7652 kHz would sign off soon afterwards as 0600 was close to sunrise in Liberia, and 7652 kHz would not be a good frequency for long distance propagation during the daytime. Sunrise in Liberia took place before midnight in California. One curious aspect to KKN44's operation on 7652 is that the Voice of America (VOA) operated a SSB "feeder" station on 7651 kHz from its Greenville, NC, facilities. The feeder station transmitted VOA programming for reception and relay by overseas transmitter sites. The VOA signal was quite strong, although KKN44 managed to be heard well through the interference. The VOA feeder station remained on the air until 0600 during the last half of 1988.

As Harry monitored KKN44, he noted how its signals were always quite strong even when signals from African shortwave broadcasters (such as the Voice of Nigeria) were marginal on the nearby 7100 to 7300 shortwave broadcasting band. This was surprising, since the information filed by the U.S. State Department with the International Telecommunications Union gave KKN44's power as only 1000 watts. It was clear from its strong, steady signals that KKN44 was using considerably more power.

While monitoring KKN44, Harry discovered that it also broke from its "canned" marker around the hour and half hour for hand-sent versions of the marker and CW groups. The surprise came at the sign off times. On October 16, 1988, Harry found KKN44 still sending its QRA marker past 0700 UTC. At that time, it was well past sunrise in Liberia, and most of Europe and Africa were in daylight yet the KKN44 signal was still strong. Harry continued monitoring through the last part of 1988, and on December 11 caught KKN44 with loud, steady signals on 7652 kHz at 0838 UTC. That was 8:38 a.m. in Liberia, and all of Africa and Europe (except for the polar regions) were in daylight. In fact, the sun had risen in eastern Brazil by that time. African shortwave broadcasters in the 7100 to 7300 kHz range had faded out over an hour earlier, yet KKN44 still was putting S9+ signals into southern California. As Harry later wrote about his reception, "unless they're repealed some of the laws of physics, I don't think the station I heard that December 11 could have possibly been in Liberia."

Another person who was monitoring KKN44 during that period was George Zeller, the editor of the "Outer Limits" column in ***Monitoring Times***. During his listening, George made the surprising discovery that the KKN44 signal was actually part of the VOA feeder signal while the VOA was on the air. In a letter to Harry, George wrote ". . . the CW was definitely on the USB part of the feeder on 7651, not off by itself as a SW signal on 7652. I checked this carefully on all modes on the NRD-525." (The NRD-525 is a high performance receiver used by George.)

Harry was convinced that the signals he heard on 7652 kHz during the fall of 1988 did not originate from Liberia, and feels today they probably came from a transmitter in the United States. He did hear one signal that he suspects may have actually come from Liberia or another African location, however. On January 7, 1989, at 0635 UTC, he noted the usual QRA from KKN44 was missing. In its place was a much weaker, hand-sent "9GV 9GV DE

KKN44 KKN44" marker. "9GV" was a callsign from the series allocated to Ghana. He had checked 7652 kHz earlier that evening prior to the VOA sign off and had noted no trace of KKN44. The next night, the usual KKN44 QRA marker was back at its normal strength.

Harry resumed monitoring KKN44 during 1990 as the civil war in Liberia broke out. He listened to KKN44 on 4886 and 7652 kHz beginning in June, and noted no variations from its usual operations despite the presence of numerous American citizens in Liberia and the U.S. Navy offshore to assist in any possible evacuation of them. However, July 31, 1990, marked the last "normal" reception of KKN44. 4886 kHz closed promptly at 0600 that date while 7652 kHz continued in operation. It was not heard again until August 4, when Harry caught KKN44 on 4886.2 kHz at a very weak level around 0400 with its QRA in hand-sent CW; nothing was heard at that time on 7652 kHz. Harry continued to check both frequencies for several days, but heard nothing else until August 12 at 0330 on 4886 kHz, when he heard random CW groups at a weak level. He heard no identification, however, and KKN44 has yet (at the time this book was being prepared) to return to the air. In a related intercept, Dave White of Maine reported hearing KKN50 on 6925.5 kHz sending "ZLN KKN44 32/41/43/69 K" at 2255 UTC on August 5.

Some SWLs also became curious about station KKN39 on 4956.7 kHz. Knowledgeable Florida DXers such as Dr. John Santosuosso noted the station could be heard loudly in southern Florida even at noon, which would be almost impossible from a propagation standpoint if KKN39 was really located at its listed Washington, DC, location. Efforts were made to discover a transmitter site for KKN39 in Florida, and the May, 1989 issue of ***Monitoring Times*** reported that a transmitter site had been located at Krome Avenue and State Road 997, 15 miles west of downtown Miami, near the Zoological Park and Gold Coast Railroad Museum. During September, 1990, that area (along was much of Dade County) was devastated by Hurricane Andrew. During the evening before Andrew made landfall, KKN39 was well heard on 4956.7 kHz. The next evening—after Andrews had passed through the Miami area—KKN39 was gone and remained off the air for several months thereafter.

It certainly seems to be true that when political events take place at overseas locations, KKN activity picks up with increased callups on the marker frequencies. But as mentioned previously, actual messages are seldom seen on the marker frequencies. One good example came on January 16, 1991, at the start of Operation Desert Storm. Dave White was monitoring KKN50 on 6925.5 kHz at 0014 UTC on that date, and he caught "KKN50 KKN50 DO YOU COPY KK?" sent then in CW. KKN50 apparently did not hear this station or else ignored it, as it continued to send its QRA marker. At 0022, KKN50 interrupted the marker and apparently started retuning its transmitter. At 0026, "KRC81 KWT94 DE KKN50 QSX 6/10/12 K" was sent and this was repeated until 0050. KKN50 abruptly went silent at that time, and "NJEEOEEE" was heard on the frequency. Dave described the signal as loud and apparently sent automatically. At 0115, Dave heard "KKN50 DE KWT94 ZHQ 14368/11424 K," "DE KKN50 INT ZHQ 14368/11424 K," and "DE KWT94 C" sent by hand. Dave checked 11424 kHz at this point and heard KKN50 there as well. On January 17, Dave heard similar traffic on 6925.5 kHz from 0014 to 0117 UTC. At 0105 on that date, Dave heard "XVWE XVWE DE RKV NCV KKK" and at 0106 "XWWE XWWE RCV RCV RCV KKK." Alerted by Dave, Harry Helms was listening to 6925.5 kHz on January 18 and caught "KWT91/KWT94 DE KKN50" being repeated at 0228. This continued until 0400, when KKN50 resumed its QRA marker.

This pattern of activity was repeated in December, 1992, when U.S. forces were dispatched to Somalia. On December 5, Dave was listening to KKN50 on 11458 kHz at 2155, when its QRA marker was interrupted by both KWT91 and KGN39 calling KKN50. Their messages were cryptic, such as "QJK 1 ZTA3? Q EEE QJK 1 ZTA3 K." At 0001 UTC (which was December 6 UTC), KKN50 began sending "KWT91 DE KKN50 ZBO O ZZK K," which continued until 0100. At that time, several random letter groups were sent in CW. At 0116, a "KWT91 DE KKN50" marker began, and this continued until 0145 when the usual QRA marker resumed.

A rare but very curious aspect is the use of modulated CW (MCW) by KKN50 and KKN39. As explained in Chapter 1, modulated CW is nothing more than Morse code sent over a conventional AM transmitter by audio tones. Harry Helms reported hearing KKN50 on 6926 kHz (not the usual 6925.5) sending its QRA marker in MCW on March 4, 1990 at 0430. The pseudonymous "Havana Moon" reported a similar use by KKN39 on 4957 kHz at 0430 on April 29, 1990.

A final interesting aspect is how the most widely heard overseas KKN stations are in nations where the Voice of America (VOA) maintains relay stations. Examples include KRH50 (Great Britain), KWS78 (Greece), and KWL90 (reportedly in the Philippines). Liberia was also the site of a major VOA relay station in addition to KKN44; the VOA station left the air a couple of days after KKN44 and neither has returned to the air as of the time this book was published. The presence of VOA relay stations and KKN stations in the same country may be a coincidence, but may be significant in light of George Zeller's observations that the KKN44 signal was present on a VOA signal apparently originating from the VOA's Greenville, NC, transmitter site.

The National Communications System (NCS)

When the Warrenton, VA, site for four-digit numbers stations was discovered, there was a bonus: the 6925.5 kHz signals from KKN50 also were originating from the same site. And, as mentioned in Chapter 1, a sign at the facility showed it was part of the National Communications System.

When the Miami transmitter site for KKN39 was located, a sign at the facility said it was also part of the NCS. In fact, the telephone number for the facility was listed in local telephone directories under "U. S. Government" as "National Communications System."

The NCS was created by an executive order of President Kennedy in 1963. The basic purpose was to coordinate the communications facilities of the U.S. government and provide a common operational framework for all situations from routine matters through national emergencies. Civilian and military systems are included in the NCS, and the Secretary of Defense has overall executive authority for NCS.

A few years ago, ***Popular Communications*** editor Tom Kneitel wrote an article for that magazine about diplomatic radio communications systems, focusing primarily on foreign embassies located in Washington. In response, Tom received a letter from a support communications officer (SCO) stationed at a U.S. embassy over-

AA 3215	AP 4890	BG 6874	BX 10896	CQ 15820
AB 3220	AQ 4920	BH 7650	BY 11036	CR 15840
AC 3237	AR 5065	BI 8062	BZ 11526	CS 14876
AD 3250	AS 5070	BJ 8186	CA 11606	CT 17630
AE 3300	AT 5090	BK 9052	CB 11692	CU 17660
AD 3980	AU 5097.5	BL 9112	CC 12186	CV 18560
AE 4061	AV 5115	BM 9206	CD 12190	CW 19470
AF 4460	AW 5207.5	BN 9284	CE 12318	CX 19490
AG 4520	AX 5250	BO 9460	CF 13392	CY 20269
AH 4540	AY 5360	BP 9470	CG 13554	CZ 20940
AI 4545	AZ 5790	BQ 9786	CH 13732	DA 22820
AJ 4580	BA 5810	BR 9888	CJ 14832	DB 23390
AK 4610	BB 5830	BS 9952	CL 14679	DC 24660
AL 4785	BC 5845	BT 10350	CM 14910	DD 26142
AM 4830	BD 5882.5	BU 10396	CN 15513	DE 27325
AN 4870	BE 6795	BV 10488	CO 15594	
AO 4885	BF 6840	BW 10698	CP 15732	

Figure 3-3: A copy of the "CS Authorized Frequency List" supplied by an anonymous U.S. State Department communications officer.

seas. This SCO took issue with some of the observations Kneitel made, mostly about the use of diplomatic pouches. Along with his letter, this SCO sent along an enigmatic frequency list entitled "CS Authorized Frequency List." The SCO made handwritten notes indicating it was used by a NCS "communications school" at the "Warrenton training center." This list is reproduced in Figure 3-3.

One interesting aspect about this list is how several of the frequencies—such as 3250, 4460, 5810, 6840, 8062, 8186, 9052, and 11526 kHz—were extensively used by numbers stations in the mid-1980s. Other frequencies were used in that period by KKN stations, such as 15594 by KRH50 in London. (Frequency "BH," 7650 kHz, is close to KKN44's 7652 kHz.) As can be seen in the Underground Frequency List, some of these frequencies, like 6840 and 8186 kHz, remain very active. However, since this list is a least seven years old, there is the possibility that a new list of frequencies is now in use.

This list has created much speculation among SWLs. For example, are the signals heard on such frequencies as 6840 and 8186 kHz actually training exercises rather than real-world communications? Other questions surround the "communications school" at the "Warrenton training center." Is it a legitimate training facility, or a cover for something else? If it is a real training center, just who is being taught and what are they being taught?

NCS facilities are scattered across the country, and some, like Miami, are even listed in local telephone directories. It is safe to assume that any U.S. government shortwave transmitting facility is part of NCS. Since both numbers transmissions and KKN50 were found to originate from the same NCS site, it is conceivable that numbers stations and other KKN outlets could use multiple NCS sites anywhere in the United States. In fact, some SWLs speculate that the NCS is the key to many mysteries, with numbers stations, KKN stations, and FEMA coded operations all falling under its jurisdiction. While hard information about NCS is still scarce, it is clear that SWLs will spend much time in the future to determine what role NCS plays in the mysteries discussed in this book.

4

Miscellaneous Mystery Signals

THERE ARE MANY OTHER unusual things you can hear on shortwave radio, and this section is a collection of some of "the real good stuff!"

"Pulsers" and "Beepers"

Some stations do nothing more than transmit various "beeps," dashes," "pulses," or "ticks" at fixed repetition rates of typically 60 to 90 per second. When first heard, these stations may sound similar to time signal transmissions. However, these stations do not identify, nor are they listed in any standard frequency lists such as those of the International Telecommunications Union. Moreover, many of these are pulses of unmodulated carrier, similar to what would result if an electronic keyer were continuously held to the dot or "dit" side. The accuracy of these transmissions can be very good; some have been observed in perfect synchronization with WWV and WWVH. Sometimes these signals are interrupted by data bursts or coded groups; it may be that these signals are "markers" of some sort to hold a channel between messages.

"Raspers"

Several stations can be heard which transmit what seem to be patterns of dots and dashes using "noise bursts." The effect is similar to trying to send Morse code by hissing it, although the combinations of dots and dashes usually form no letter found in any version of the Morse code. For lack of a better name, these stations have become known as *raspers* because of the rough, harsh sound of the transmitted dots and dashes.

These signals have been around sine the 1970s and were rumored to be connected with U.S. military operations. Several observers who were in the Gulf area during the Iraqi conflict commented they heard many raspy dots and dashes signals through the HF bands and on up into the VHF bands. It would certainly appear, therefore, that these transmissions are definitely associated with U.S. military activities.

The AN/USQ-76(V) equipment is believed to be the source of this signal. Manufactured by Magnavox, it is a digital data terminal set which provides control functions with either HF or UHF radio equipment. It is appropriate for use at ground stations as well as on board ships.

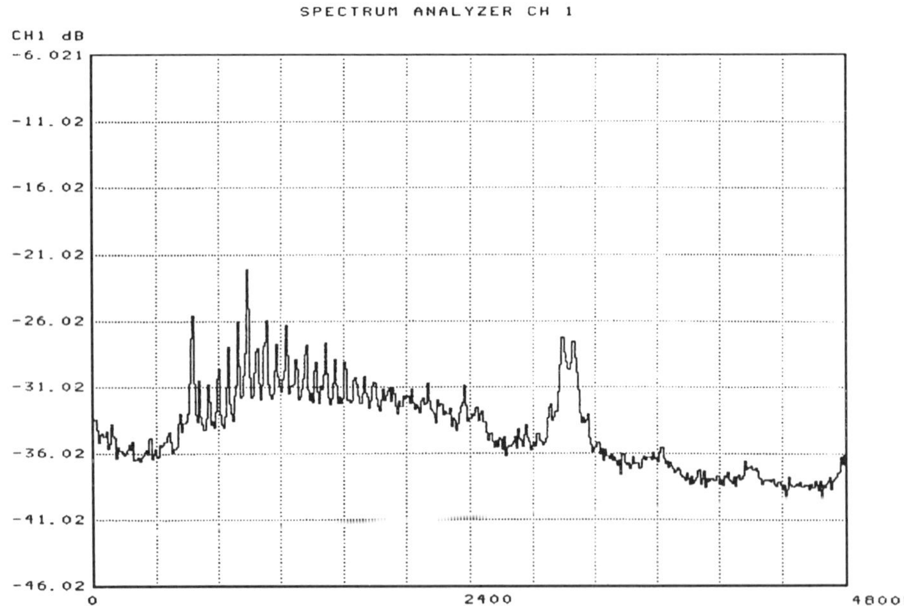

Figure 4-1: Kevin Tubbs of Vermont used a spectrum analyzer to display a "rasper" signal he found on 11186 kHz. Here is the chart of the signal's spectrum.

The pattern of dots and dashes in these transmissions can change within a few hours on the same frequency. As of yet, no one has come up with a good explanation of why the patterns change or what the patterns signify. Figure 4-1 shows a spectrum analyzer display of one of these signals. This particular display was provided by Kevin Tubbs of Vermont.

Raspers seem like the American equivalent of SLHFMs—a lot of widely heard activity without any readily apparent meaning. More monitoring is required to solve this mystery!

The "Foghorn"

Beginning in late 1989, a new signal that became known as the "Foghorn" was heard throughout the shortwave bands. The Foghorn sounds like a loud "BBRRRAAPP!!!!" The usual pattern is for it to operate in brief "bursts" of five to thirty seconds followed by silence for approximately a minute. The signal will often change pitch or tone during its "on" period, and signals are usually quite loud; it's not uncommon for a Foghorn to be the loudest signal on a band. Sometimes multiple frequencies will be active with Foghorn signals. Some SWLs feel there are two types of Foghorn signals that should be considered as separate signals in their own right. One is the "Motorboat," in which the pattern is repeated so slowly that the signal sounds like an outboard motor (a "putt-putt-putt" sound). The other is the "Hummer," in which the pattern is repeated so rapidly that the effect is like a raucous hum. Sometimes these signals will metamorphosize into each other; a Motorboat might "speed up" into a Foghorn and vice-versa.

The Foghorn and its relatives were broad-band signals, often occupying over 30 kHz of frequency space. The signal would be strongest at a center frequency, gradually weakening as one tuned away on either side of the center frequency. However, the decline in signal strength was gradual. A Foghorn might be 40 dB over S9 at its center frequency and drop off to S7 at the ends of the frequency space it occupies. Foghorns would often operate for a few minutes on one center frequency, stop, and then appear on another center frequency 100 kHz or so away. There was numerous cases where several Foghorn signals would be simultaneously operating within the same 300 to 400 kHz frequency range. The signal bandwidth seemed to be inverse proportion to the pattern repetition rate. Motorboats occupied the most frequency space, followed by Foghorns, and finally Hummers having the narrowest bandwidth.

During 1990, and 1991 the Foghorn was especially active. At night, it was reported on frequencies below 5 MHz while during the day it was observed over 26 MHz,

A listener in Canada monitored the signals, and he believes the pulse rates of 25 and 50 pulses per second are what produce the Foghorn sound. It is believed these signals are emitted by U.S. over-the-horizon-radar (OTH). The July, 1990 issue of *IEEE Spectrum* carried an extensive article on U.S. OTH systems, and the signals described in that article did closely approximate the reported Foghorn, Motorboat, and Hummer signals. Both the U.S. Air Force and Navy have this type equipment. Bearings obtained by Canadian SWLs indicated some Foghorn transmissions were possibly originating at the Whidbey Island Naval Air Station in the state of Washington.

Monitoring Times columnist George Zeller discovered a General Accounting Office (GAO) publication entitled *Over the Horizon Radar: Better Justification Needed for DOD Systems' Expansion* (document GAO/NSIAD-91-61). This was a report prepared by the GAO in 1991 to assess U.S. OTH programs for Congress. The report stated that both the U.S. Navy and Air Force had been testing OTH systems since 1989 from sites in Virginia, Alaska, Idaho, California, and Oregon. In addition, the report said the Stanford Research Institute had been testing another OTH system from a California site beginning in 1990. These dates correspond with the onset of Foghorn activity, and the various sites around the country would explain the powerful, local-like signals note by some SWLs.

At the time this book was being prepared in mid-1994, the overall level of Foghorn activity heard in the United States was down from its peak in 1990 and 1991. This may be related to cutbacks in the U.S. defense budget. However, European monitors still frequently report these signals. Whether the European signals originate from U.S. bases in Europe or from European military forces is not known at this time.

Russian Spaceships and Tracking

When the satellite age began with the launch of Sputnik I in 1957, owners of shortwave radios could tune in the "beep-beep-beep" telemetry signal is transmitted near 20 MHz. Despite all the technical advances since then, the Russian space program still makes use of frequencies near 20 MHz for its manned space missions.

The most active channel is 19954 kHz, used for beacons and telemetry by numerous Russian space vehicles. It is mainly used by the Mir space station (actually, the beacons are on the "Progress" resupply vehicles docked to Mir). These signals consist of a "beep" or pulse at intervals of approximately one per second, and interrupted every 45 to 90 seconds by a "trilling" or "burring" sound. This sound is actually a data transmission. Signals actually coming from space are easy to identify due to Doppler shift. Doppler shift is a change in the apparent frequency of a signal due to motion. This is the same effect that causes the pitch of an ambulance siren or train whistle to vary as the vehicle moves toward and then away from you. With Mir, the signal will first appear to be higher than 19954 kHz. As Mir approaches your terrestrial receiving location, the signal will appear to steadily drop and will be nearest 19954 kHz when Mir is closest to you. As Mir moves away from you, the signal will seem to drop below 19954 kHz. To best receive Mir, use the USB mode and widest bandwidth of your receiver. Since Mir is audible for only a few minutes during each "pass," and Mir crosses a different area of the Earth on each orbit, you may need to extensively monitor 19954 kHz before you hear Mir. However, the signals from Mir are usually quite strong.

20008 kHz is often used by Soyuz manned vehicles from launch until docking with Mir and during the return to Earth phases of manned missions. Both CW and AM voice are used, and signals from Soyuz are identifiable by the Doppler shift.

Instead of land-based tracking stations, Russia has an extensive fleet of tracking ships named after various cosmonauts and famous Russian scientists. These stations communicate using CW and RTTY on several network frequencies which are active during Russian manned space missions.

Chirpsounders

Many of you have probably been monitoring a frequency and heard a chirpy signal go sweeping by. If you follow the signal, you discover that it sweeps from 2 to 30 MHz, requiring about five minutes to make a complete sweep. This signal is generated by a Tactical Frequency Management System, the AN/TRQ-35 (V) manufactured by BR Communications of Sunnyvale, California. The system was designed to provide an operator with a means of checking for the best frequency for communications under changing conditions. There were three primary components in the first generation equipment: a spectrum monitor, a chirpsounder transmitter, and chirpsounder receiver. There is widespread use of this system with the equipment in service with some NATO countries, the U.S. Air Force, Navy, Marines, and Army, and the British Navy and Air Force.

The second generation of equipment was more compact and provided additional features including enhanced performance. The upgraded model had a FMT-5A Frequency Management Terminal in addition to the previously indicated three units. There was also available a RCS-5B Frequency Management Receiver which incorporates the spectrum monitoring capability and the functions of the RCS-5 Chirpsounder receiver. As was the case with the RCS-5, the 5B is capable of operation as a synthesized SSB HF communications receiver. The HF operating parameters measured are received power, multipath, and propagating information.

Piccolo

This type of signal is properly named because it does resemble the sound of the musical instrument. Others have described it as a bubbling (like the old Maxwell House coffee commercials) or "tinkling" sound. The system has been in use by the British for some time with the Foreign Office and Diplomatic Service as major users. Other users are the Commonwealth Office, the British Defense Communications Network, and the British Army. There are other variations of the piccolo system in use by some European nations, and there was also a Russian piccolo system reported being used in the past. Decoding equipment for piccolo is complex compared to that required for RTTY or other digital modes. The July, 1989 ***Monitoring Times*** carried a report on the first successful decoding of piccolo signals by a radio hobbyist.

"Water Drippers"

According to the information available at this point in time, there appear to be at least two types of "water dripper" signals. The one heard on 8040 kHz is reportedly over-the-horizon-radar located at Halifax, NS, Canada. Another type logged sounded like a series of rapid and irregular drips. It is speculated that this latter type could be a multi-frequency step-tone data transmission.

Smuggler and "Bootlegger" Communications

The 1980s saw the introduction of versatile new amateur radio transceivers with general coverage receiver sections tuning (typically) 150 kHz to 30 MHz. It so happens that these transceivers can also transmit anywhere in that frequency range with amazingly simple modifications. For example, one popular transceiver by Kenwood can be converted to "general coverage" transmit operation by removing two diodes from a circuit board. Another transceiver, from Yaesu, merely requires moving an internal switch to the opposite position for transmit capabilities from 150 kHz to 30 MHz in AM, FM, CW, and SSB modes! Couple these impressive capabilities with low price and easy availability, and it's no surprise that all sorts of unusual communications networks have sprung up recently.

Drug smugglers have been very quick to use this new communications technology to coordinate the shipment of narcotics. In fact, the Drug Enforcement Administration reportedly devotes considerable effort to trying to track down smuggling communications on shortwave. Most smuggling communications are in Spanish, English, or a mixture of both. As you might imagine, these nets are not formal, and radio procedure is sloppy. Whistling, sound effects, and endless "holas" are used to attract attention and establish contact. You will seldom hear an explicit reference to drugs; many code words, some quite obscure, will be used. For example, some monitors have reported hearing people talking about a load of "brownies" or "packages" or "the shipment." Much of the information will be about rendezvous points, meeting times, schedules, etc., all presented in somewhat cryptic and elliptical terms.

Not all informal nets will be drug smugglers. Commercial fishing fleets have also set up extralegal nets on the shortwave bands, and these are rapidly increasing. At "first listen," you might suspect these are also smugglers. However, much of the conversation relates directly to fishing and related topics, usually sex and drinking. If your ears are easily offended by profanity, you'll quickly tune away from these nets!

I recall that while monitoring one evening I heard two individuals discussing a third individual. It was difficult to determine exactly what the third person was involved in, but I remember hearing him described as seeing something or someone on the dock. He backed out of his dock position at 100 miles per hour. Now, that is pretty fast for a boat going backwards. Anyway, I gathered that he had cleared the area rapidly. Maybe he was a smuggler and had detected law enforcement personnel on the dock, or maybe it was simply a case of him owing money to someone and he saw the person on the dock and thought it best to get out of there??

Once, when copying fishing boat communications, I noted a boat near Panama talking to a boat apparently in Alaskan waters. One individual was complaining about the heat, while the other was complaining about the ice forming on his boat deck.

The "freebanders" who have traditionally operated between the CB band and the 10-meter band have expanded their horizons. These transmissions sound much like a cross between ham radio and CB operations, except that bogus call signs (such as "Unit 14-22B") are used. Even long-haul truckers are reportedly getting into the act from 6700 to 7000 kHz!

Finally, high-tech bootlegging has arrived with the use of packet radio. Packet radio is a system whereby personal computers can be linked by radio to automatically store and forward messages and request retransmission of messages that were not correctly received. The effect is similar to RTTY, except that completely unattended operation is possible. Networks like this were discovered in late 1989 operating on shortwave and the use of packet has spread rapidly since then.

"Water Gurglers" and Spread Spectrum

Another type of signal sometimes heard on shortwave sounds much like gurgling water. Like the Foghorn, these signals occupy considerable bandwidth (often over 10 kHz) but are nowhere near as strong as the Foghorn. These signals are believed to be a form of spread spectrum communications. Spread spectrum is a modulation technique where the transmitter power is spread over a wide frequency range rather than being packed into a narrow range, as is done with conventional AM, CW, or SSB techniques. One method of spreading the spectrum of a signal is to rapidly "hop" the carrier about a frequency range in accordance with a "spreading code." The receiver also has a copy of this spreading code so it can "track" the carrier within the bandwidth. The advantages of spread spectrum on shortwave include communications security (the signal cannot be demodulated without the appropriate spreading code) and a high degree of immunity to jamming because the spread spectrum receiver ignores signals that are not spread according to the code it is programmed to recognize. A ham attending a defense electronics show saw a demonstration of the AN/PRC-104A "anti-jam" short-wave transceiver, which uses spread spectrum technology. Part of the demonstration included hearing what the transmitted signal sounds like on a conventional receiver, and the ham reported the result was the "water gurgler" sound. It is likely that more and more spread spectrum signals will be heard on shortwave in the years ahead.

The "Mystery Translator"

A *translator* is a circuit that takes a certain frequency range of received signals and then retransmits them on a separate frequency range. For example, a translator might receive the 7000 to 7015 kHz range and retransmit it on the 8650 to 8665 kHz range. All signals falling in the original 7000-7015 kHz range will be retransmitted, including CW, SSB, AM, and even static crashes. Translators do this by taking the received frequency range, mixing it with a local oscillator, and them amplifying the resulting frequency range. Translators are commonly found in the VHF and UHF ranges, often as part of a system to extend the coverage of FM broadcast stations or in amateur radio communications satellites. However, at least one translator system has been observed in operation on the shortwave bands.

An extensive report on this translator appeared in the June, 1992 issue of ***Popular Communications*** magazine. The author, Dwight Brown, is the operator of amateur radio station W5WE. He had first observed the translator in operation in 1989, and his most recent receptions were in December, 1991. The translator appeared to cover the range of 7013 to 7015 kHz, with an output range of 8878 to 8880 kHz. The signals on the "translated" range were strong, indicating that it must use high transmitter power and a good antenna system.

Dwight enlisted the aid of several fellow hams to monitor the translator output, and found most of the ham signals it retransmitted originated in Russia, although signals were also noted from hams in Britain, the United States, Canada, Yugoslavia, Sweden, Germany, France, Spain, and Italy. The signals on 8878 to 8880 kHz were mainly audible in the eastern half of North America during December, 1991 from sunset to about 0200 UTC. Giving the propagation and that time and the locations of the signals being retransmitted, Dwight felt the most likely location for the system was somewhere in eastern Europe.

On December 10, Dwight went on the air as W5WE and contacted Virgil Holobaugh, W5LMS, for the purpose of trying to access the translator. Dwight transmitted on 7015 kHz while Virgil listened on 8880 kHz. By telephone, Dwight was able to hear his "translated" signal as received by W5LMS on 8880 kHz. Dwight heard a slight delay in his received signal over the phone, a sign his signals were being relayed.

Dwight ran a similar test on December 15 and contacted a FCC monitoring station about what he was hearing. The engineer taking his call suggested he was hearing a receiver image. After tuning the range for themselves, however, personnel at the FCC monitoring station Dwight contacted expressed surprise at what they were hearing. While no connection can be proven, the translator was off the air the next day and has not been heard since. The purpose and location of this system remains a mystery. However, should you run across a narrow range of signals that are not where they're supposed to be (such as ham signals in the middle of an aeronautical band), be alert to the possibility that the "mystery translator" may have returned to the air!

A VOA/Mystery Station Link?

The June, 1990 issue of *Clandestine Calling* (published by Harald Kuhl, Weender Str. 30, D-3400 Goettingen, Germany) told of an unusual signal then being heard in Europe on 20125 kHz in USB. The station seemed to operate continuously with music and scrambled speech. A breakthrough in identifying the station took place on May 24, 1990, when Sweden's Claes Olsson reported hearing the music stop and the following male voice traffic in the clear: "Greenville calling. Greenville calling Liberia. Do you copy us?....Gary, Gary Wise...Liberia, Liberia, Greenville calling...over." There was a pause, and people could be heard speaking in the background before the male voice traffic resumed. "Hello Liberia, hello Liberia. . . . Liberia, this is Greenville calling, this is Greenville calling, do you read us? Over." After this, the music resumed.

It so happens that 20125 kHz was listed for use by a VOA feeder station at Greenville, NC. This was the same VOA transmitter site where the signal of KKN44, supposedly at the U.S. embassy in Monrovia, Liberia, was found on a 7651 kHz VOA feeder channel. One big question raised by this reception is why scrambled voice traffic was transmitted over a station ostensibly used just to relay VOA programming to its Liberian relay station.

Air Force One Communications

Air Force One uses several shortwave frequencies for communication while aloft, and some of the more active ones are given in the Underground Frequency List. Air Force Two also operates of many of those channels. The identifiers "Air Force One" and "Air Force Two" are used only when the president or vice-president are aboard the aircraft. If the president or vice president is not aboard their respective aircraft, the identifier will begin with "SAM," which stands for "special air mission." Some SWLs have heard the identifier change from a "SAM" one to "Air Force One" when the president boards the plane at Andrews Air Force Base or some other location.

Most transmissions from Air Force One will be routine aeronautical communications relating to air traffic control, arrival times at a destination, etc. Sometimes the shortwave channels will be used for phone patches or other sensitive communications, and these will be normally be scrambled. However, on rare occasions some important communications, including telephone calls, inadvertently go out "in the clear." This often happens in the early weeks of a new administration, as was the case when President Clinton took office, apparently because some members of the new administration forgot to activate the scrambling devices before making telephone calls. For example, SWLs were able to hear phone calls relating to appointments in the new administration, including candid assessments of the strengths and weaknesses of potential appointees. The same thing happened during the Bush and Reagan administration. In one case, "Rawhide" (the Secret Service code name for President Reagan) was heard discussing plans with the Secretary of Defense for possible military retaliation against suspected terrorists.

"Non-Covert" Communications

Much of the stuff that sounds mysterious on the shortwave bands really isn't covert or underground radio. While a transmission like "Bomber, this is Dispel, QSY alpha five" certainly sounds like covert traffic, it is actually the sort of thing you can regularly hear on military communications channels. The armed forces of the world often use tactical callsigns (such as Missionary, Twin Pod, Pinstripe, etc.) as well as coded voice traffic. Unlike true covert communications, military traffic is characterized by tight discipline and operating procedure, and fixed channels are frequently used. For example, 8967 kHz is often used worldwide by the U.S. Air Force while 14492 kHz is used for ship-to-ship communications by the Russian navy. Most such communications will be in USB, and voice scrambling is common. Some law enforcement agencies of the U.S. government, such as the Customs Service and Drug Enforcement Administration, also use tactical callsigns and coded messages for their operations.

These military and government communications are beyond the scope of this book. To learn more about them, consult a good basic reference like the ***Grove Shortwave Directory*** (P.O. Box 98, Brasstown, NC, 28902-0098). With experience, you will be able to quickly determine whether a signal is truly covert or tactical military or government communications.

5

The Underground Frequency List

HERE IS A LIST OF COVERT, unidentified, or unusual signals reported in the two years prior to publication of this book. By the time you read this book, not all of these signals will still be heard, and new unusual signals not in this list will be heard in their place. However, this list is indicative of the scope and breadth of covert and unusual signals that you can hear on shortwave. Several of the frequencies in this list—such as for FEMA, Air Force One, the KKN stations, and numbers stations such as 6840 and 8186 kHz—have been active for several years and will likely be still in use. The word **active!** has been used to note frequencies especially busy during the previous two years. If you can't hear a signal at the time or on the frequency indicated in this list, try tuning a few kHz above and below the listed frequency and listening an hour earlier or later than indicated.

A dot (•) is used to indicate a separate signal is heard on the same frequency. Sometimes the difference between signals is obvious (such as a different language or modulation method), while in other cases is more subtle, such as a difference in message structure and organization. It may be in some cases that multiple signals on the same frequency have a common source (even down to the same transmitter site).

Note how several frequencies (such as 7541, 7655, 7887, 9251, 10125, and 13890 kHz) are "home" to several different types of unusual signals. It is tempting to speculate that all such signals on a given frequency have a common source, but other factors (propagation, lack of interference, etc.) may account for the presence of several mystery signals on a single frequency. In your own monitoring, be alert for such patterns.

All times in the following list are in coordinated universal time (UTC), also known as GMT. The following abbreviations are used in the list:

CC	Chinese
EE	English
GG	German
JJ	Japanese
RR	Russian
SS	Spanish
CW	Morse code (unmodulated)
MCW	Tone-modulated Morse code
AM	Amplitude modulation
SSB	Single sideband
USB	Upper sideband
LSB	Lower sideband
ISB	Independent sideband
RTTY	Radioteletype
YL	Female announcer
OM	Male announcer
3-2F	Three-digit group, pause two-digit group
5L	Five-letter groups
5C	Five character (letters and numbers) groups
4L	Four-letter groups
5F	Numbers in groups of five (including cut numbers)
4F	Numbers in groups of four (including cut numbers)
MFA	ministry of foreign affairs/department of state
EMBACUBA	Cuban embassy (header in message traffic)

A special note: there seems to be a bit of confusion on the part of some monitors in properly assigned the callsign heard in a call-up. For example, if we hear a station sending "DE XYZ XYZ XYZ," we generally accept XYZ as being the transmitting station. However, when we see a call-up such as "XYZ, XYZ, XYZ," it is most likely that XYZ is the callsign of the station being called, the one who the traffic is intended for. This convention has been followed in the following list. However, I have noted monitors frequently identifying the callsign in this latter case (XYZ) as being the transmitter callsign.

1620 "SBT" repeated continuously in CW approximately 0200 to 1000; believed to be in Cuba.

1646 OM and YL in what seems to be French around 0515; audio highly distorted and may be some form of narrow-band FM. Best heard in eastern United States and believed to be from either Haiti or possibly St. Pierre et Miquelon.

2180 EE/YL in AM with 5F groups 0400.

2200 YL/SS in USB with 4F groups 2200.

2270 YL/EE Mossad station "Juliet Sierra Romeo" at 2000.

2320 FEMA USB operations.

2331.5 OM/JJ network in USB at 0546; believed to be fishing vessels.

2360 FEMA USB operations.

2377 FEMA USB operations.

2445 FEMA USB operations.

2651 Two dots, one dash rasper 0931.

2738 OM/EE traffic in USB 2358 at weak level; had Jamaican accents.

2771 Three dots, pause, 12 dots, one dash rasper 0845.

2800 Unidentified scrambled speech communications in USB at 2016. Also heard on 3003, 3020, 3030, 3567.5, 3480, 3920, and 3950 on various days and at various times.

2870 "NR 83364 NR 6U NR 83364 NR 6U QRU QRU" in CW 0219.

2926 OM/SS network in AM 1100-1300; sounds like a CB channel and is believed to be in Sonora/Baja California states of Mexico.

2955 CW network around 2130 to 1100 with calls "GCA," "WHY, "AVO," "IGK," and "INU" as net control. Hand-sent CW; believed to be Cuban Border Guards.

3006 OM/EE fishing net in USB 0517; X-rated conversation.

3100 Three dots, one dash rasper 0230.

3109 "T5M" working "E6V" in USB 0515; British accents.

3150 YL/EE announcing "Papa Charlie Delta Two" in AM at 2301.

3175 5L groups in CW 0210; includes "ñ" character.

3177 "SVK" acting as control station for CW network around 0000. Other calls heard included "GXC," "MKH," "XRA," etc. Traffic was mixed letters and figures.

3183.2 "4NH DE 6HA K" in CW 0151.

3205 CW message in groups of 5 characters which included letters A-Z, numbers 1-0 and some punctuation marks plus four special characters. Heard at 2319 and still sending at 0306. Also heard on 3606 kHz.

3255 CW message with unusual characters (see 3205 kHz entry) at 2326, parallel to 3637 kHz. Stopped at 2350.

3262 YL/GG repeating "Romeo Delta" with electronic tones from 2000-2005. 5F traffic was for "041/12," "116/40," "208/33," and "457/24." Reported next day at 1900 on 4543 kHz.

3280 Czech station OLX at 2100 with "597" in CW call-up. Also on 5301 and 8142 kHz.

3292 YL/SS in AM with 5F groups 0403.

3302 Unidentified station sending messages in CW at 2320. Groups have numbers and usual characters. Dualing with 3700 kHz, still going at 0300.

3341 FEMA USB operations.

3378 WGY912, Mt. Weather, VA, with 3-character groups in CW at 0206.

3382 Three dots, one dash rasper 0634.

3388 FEMA USB operations.

3397 OM/SS in USB 1140 with military-sounding traffic and number groups; believed to be Mexican army.

3410 YL/EE in AM with 5F groups 0403.

3417 YL/EE in AM with 5L groups from phonetic alphabet at 0220.

3419.5 "BTO" repeatedly in hand-sent CW 0031, then "QAP CFM AA AR" and off.

3429.8 OM/SS traffic in USB 0031 sending dinomes to each other. Frequents breaks with "digame" (tell me) and "adelante" (go ahead).

3444 Scrambled speech in USB 2056.
• YL/SS in AM with 5F groups 0200.

3496 Scrambled speech in USB 0232.

3499 One "ping" sound per second in USB 0655; sounded like sonar.

3592 SLHFM "P" 2216, parallel to 4043.2 kHz.

3772.8 CW pulses at rate of 90 per second 1230.

3775 5F CW message with unusual characters at 2006.

3807 SLHFM "P" in CW at 2122 parallel to 4043 kHz. At 2125 RTTY started (75 baud). Messages were in 5L groups. At 2130 "P" marker took over again. At 2159 CW "P" marker followed by RTTY (50 baud) 5L group messages at 2215. At 2229 "P" marker again.

3830 Strange two-tone beeping signal at 2327. Signal had 15 kHz bandwidth from 3830 to 3845 kHz.

3861 5F CW message with unusual characters at 2138. Stopped with no signdown at 2259.

3875 "53E1" and a tone repeated in CW at 0002.

3916 Tone at 2334 steady for approximately 3 seconds followed by 10 to 15 quick dashes. Sounded like the space carrier of a FSK signal when keying quick dashes, but it was not FSK. Very strange.

3926 YL/SS in AM with 5F groups 0530.

3933 "FS4G" repeated in CW 0058.
- "ZOU4" repeated in CW 0049.
- "YKJM" repeated in CW followed by 5L groups 2225.

3961 5F CW message with unusual characters at 2355. Noted also on next night at 2347.

4020 YL/SS in AM with 5F groups 0500.

4029 YL/SS in AM repeating "atención 696 04" 0500, into 5F groups; parallel to 5762 kHz. Other 5F transmissions reported at 0600 Mondays.

4030 YL/SS in USB 0515 with 4F groups until 0520.

4032 Unidentified station in MCW at 0406 with 5F groups. Ended with "AR," sent "GR 30" and repeated message.
• YL/SS in AM with 5F groups 0130.

4035.2 "XNZ8 DE 8MNR" repeated in CW 0108.

4105.8 "MHG DE SCD QSA 5, DE RAX QSA IMI" in CW 0912.

4182.8 "UWS" repeated in CW 1315, apparently from multiple transmitter sites; clearly separate signals, keying patterns, CW pitch, and even frequencies slightly different from each other. Each site sent call once in a rotating pattern.

4192 "DRV DE CMT" in hand-sent CW 0224; "CMT" has very poor sending.

4324.2 SLHFM "R" heard at 0315.

4395 Every Wednesday at 2200, YL/GG repeating "255 255 255 81915 033" followed by five slow tones and into 5F groups at 2205.
• Another day at 2100, CW station sending "551" for 5 minutes after which "040" was sent and then into very slow CW 5F groups. Used cut numbers AUV4ENBD9T = 1-0. Also heard on 5315 kHz so it appears to be CW version of a YL/GG station.

4420 "Dripping water" sound at 2126. Signal also occupied 4600 to 4710 kHz.

4440 "P7X" in CW with message header "QRA DE P7X II P II 270300Z GR 120 BT" and then into text.

4476 YL/SS in AM with 5F groups 0530.

4495 YL/EE repeating "127" and 1 to 0 count, then "105" and into 5F groups 0000.

4504	"FJ42/59" in CW with "VVV" marker 0014.
4535	OM/EE fishing vessels with X-rated language USB 1905.
4550	YL/EE repeating "496" with a 1-0 count, then "80" and into 5F groups 0000.
4601	YL/SS in AM with 5F groups at 0600.
4603	SLHFM "P" at 2010 with unusual format; letter P sent in pairs, i.e., "P"—then a one second pause—"P," "P"—two second pause—"P," "P"—one second pause—"P," etc.
4604	SLHFM "P," Kaliningrad. RTTY heard but too weak to read. Heard at 1439.
4605.9	5F hand-sent CW groups 0053.
4610	YL/EE in AM 0404 with 5F groups, each sent twice.
4625	"Buzzer" in AM at 0159 with one continuous two-tone buzz. Stopped and went back to regular buzz at top of hour.
4640	YL/SS in USB with 4F groups 0000; sometimes parallel to 7422 kHz.
4641	YL/EE with 1-0 count and "545" call-up at 0000 on Tuesday. At 0010 no tones, just carrier and then 4F groups in SS heard faintly in background, parallel to 5045 kHz. On another Tuesday heard YL/EE repeating "609 609 609" and 1-0 count from 0000-0010 followed by 10 beeps, "count 132, count 132," and into 3-2F groups. Repeated and then off 0035. "497" repeated with 1-0 count and abruptly off 0000.
4660	"Dripping water" sound; signal has wide bandwidth. Noted covering 4625 to 4710 kHz. Previously noted same type signal around 6750 kHz.
4665	YL repeating "Victor Lima Bravo Two" in AM 0250.

4704 One dot, one dash rasper 1023.
- Eleven dots, one dash rasper 1038.
- Five dots, one dash rasper 0230.

4718 Eleven dots, one dash rasper 0446.
- Three dots, one dash rasper 2313.
- Seven dots, one dash 0638.
- Five dots, one dash 0230.

4725 OM/EE in AM with 4F groups 0310.

4737 OM/EE fishing vessel network in USB 0545 with X-rated topics.

4779 Music box "Swedish Rhapsody" tune in AM at 2102, then YL/GG with 5F groups.

4780 "Kilo Papa Alfa Two" repeated by YL in AM 0215.
- FEMA CW operations.

4801 One dot, one dash rasper 0450.
- Nine dots, one dash rasper 1025.

4821 YL repeating "Golf Kilo" from 2130-2135. Then GG 5F messages for "856," "571," and "316." These addressees were formerly used by DFC37/DFD21 (East German station) which closed down in December 1992. Former recipients have transferred to other two-letter callsign stations.

4835 Three dots, one dash rasper 0314.

4880 At 1900 YL/EE with call-up of "Uniform Lima X-ray" at a very fast delivery. When YL said "message message, group 106 group 106" her voice went very deep and slow, but speeded up when the 5L text was sent. "Uniform Lima X-ray" and "Juliet Sierra Romeo" call-ups both on this frequency at 1800. Later "Juliet Sierra Romeo" call-up stopped and message sent to "Uniform Lima X-ray."

4882 YL repeating "Uniform Lima X-ray" from 1900-1903 then into usual 5L text. This YL not the usual one that says "November," etc.; it was a different voice. YL signed off with "end of message, end of transmission." At 2000 on 4880 kHz, usual YL up with "Uniform Lima X-ray" call-up and went into different 5L message.

5000 YL repeating "Yankee Hotel Foxtrot" mixing with WWV 0145.

5015 YL repeating "Romeo Delta" from 2130-2135 followed by GG 5F group messages for "914" and "208."
• YL/GG in USB with 5F groups 0036.

5027 SS traffic in USB 1313; sounds like a phone patch with dialing tones at start.

5030 YL/EE in AM with 3-2F groups 0615, parallel to 0615.

5045 YL/EE in AM with 5F groups 0011.
• YL/EE in AM with 3-2F groups 0017; sometimes parallel to 4640 kHz.

5091 YL/EE in AM with 5F groups 0220.
• "Charlie India Oscar Two" repeated by YL in AM 2248.

5096 Unidentified "479" in CW at 1945 with 5F groups.

5144 YL/SS in AM with 5F groups 0500.

5177 CW station sending "NNN" from 2100-2105 followed by YL/GG sending "gruppe 25" and into 5F groups.

5182 YL/GG in USB at 0149 with 3-2F groups.

5205 YL/EE in AM repeating "5316 9655 8788" at 0034. Stopped at 0039 and carrier went off shortly after.
• YL/EE in AM with 4F groups 0035.

5211 WGY933, Pikesville, MD working WGY912, Special Facility, Berryville, VA 0010.
• "O4L" working WGY904, FEMA, .Atlanta, GA, in USB at 1729.

5230 YL repeating "Mike India Whisky Two" from 2115-2120. Also on 8641 kHz. Mossad activity.

5235 YL/EE sending "32274" from 2110-2115. then "ready, ready, 20, 20" and into 5F groups.

5264 "4AN4 V4VD AN4V TE6E" repeated in MCW 0040; parallel to 6792 kHz.

5297 YL/GG in AM with 3-2F groups 0437.

5299 "VVV DE 5J42/59 K" repeated in CW at 0050.

5301 OLX, Prague, Czechoslovakia with 5F message in CW at 0000. OLX now back on original frequencies of 5301, 6758 and 8142 kHz. On another day heard at 0155 with CW marker of "VVV DE OLX." At 0200 YL/Czech in AM with 5F groups.
• YL/RR in AM with 5F groups 0201.

5303 5F CW groups 0254; off at 0258.

5305 SLHFMs "S" and "C" heard at 1736. On another day heard "P" at 2110 with "PPP AS PPP AS PPP AS" etc. (AS is CW prosign for "wait.") SLHFM "C" was audible in background.

5305.5 SLHFMs "S" heard at 1600; "P," "C," "D," at 2000; "P" at 0300.

5306.4 SLHFM "F" heard at 1210.

5311 YL/Serbo-Croat (Bulgarian Betty) in AM with "555 555 555 616 616 616 05" from 1355-1400. Then "72066" repeated and off. She is on every day for seven minutes.

5315 Every Monday at 1600 YL/GG repeating "507 507 507 48649 025" until 1605. Then five tones and into 5F groups; also on 6708 kHz. Every Wednesday YL/GG with "453 453 453 08712 027" from 1900-1905. After five tones into 5F groups, parallel to 7830 kHz.

5317 Bootleg boat traffic in USB 0251, seems to be on the Great Lakes.

5340 CW station sending "LO LO/LO LO/LO LO" from 2000-2005 then "56433" and into 5F groups.

5376.6 "95 DE 70" and similar calls with OM/SS and YL/SS operators in USB 0102.

5358 5L CW groups 0052, parallel to 7550; bad hum.

5402 FEMA USB operations.

5415 YL/SS in AM with 5F groups 0500.
• YL/EE in AM with 3-2F groups 0230, "warble" jammer in background.

5417 "UWM TG" CW call-up, then "ANW TG" and into cut numbers 0931.

5425 CW station sending "QRA QRA QRA DE KRH50," U.S. embassy, London, England at 1940.

5437 YL repeating "Papa Romeo Tango Two " in AM 0400; once heard interferring with Air Force Two in USB on this frequency!
• "Alfa November Romeo Two" repeated by YL in AM 0415.

5470 YL/EE in AM with 5F groups at 0204

5489 Unidentified CW station with 5L groups 0151.

5500 YL/EE repeating "288 oblique zero zero" 2000 to 2005; ended with "out."

5529 YL repeating "Bravo Alpha Yankee One" at 2000.

5530 YL in unidentified language every day except Sunday at 2000-2005. At 2000 repeats something like "LEHTI ASEM CINKO" which changes at 2004 to "LU BAVAM LU." Then a long tone is sent and the entire sequence is repeated.

5562 OM/SS in USB at 1000 calling "oyeron cayo largo 5562." Mentions of "Havana," "cocaine," and "300 kilos" heard.

5592 5L CW groups 0013; were hand-sent.

5610 OM/EE fishing vessel traffic in LSB 0447; lots of profanity. **Active!**

5617 Three musical notes rising in tone followed by "achtung, achtung," "31182 31182" and into 5F groups in GG by YL Sundays at 2210. Then "ende, ende" "achtung, achtung," and then repeated.

5629 YL/EE Mossad station repeating"Sierra Yankee November Six" at 0029, stopped at 0035. Another time heard YL repeating "Sierra Yankee November Two" at 1733 and also heard call-up to "Sierra Tankee November Whisky" at 2034. All USB mode.

5682 YL/SS in AM with 5F groups 0500.

5690 Caribbean-accent OM/EE network in USB 2345; references to suspects, drugs, etc. Possibly some sort of Caribbean police network.

5710 Clandestine broadcasting station and jammer at 0212 playing cat and mouse on 5710, 5720 and 5730 kHz. Clandestine station would switch frequencies and jammer would follow. Also heard another separate jammer on 5728 kHz.

5715 YL repeating "Zulu Whiskey Lima Three" at 2100.

5718 One dot, one dash rasper 0647.

5732 YL repeating "Papa Zulu" from 1900-1905 with electronic tones, then 5F groups in GG for "411." This addressee was used by DFC37/DFD21 (an East German station) until it closed early 1993. Also heard YL/GG in USB at 2308 with 3-2F groups.

5746 "Lincolnshire Poacher" tune and 5F group repeated, then into YL/EE with 5F groups in USB daily at 2100 and 2200. This station reportedly located on Cyprus and transmitting messages to Middle East. "Warble" jammers cause QRM at times. This frequency is parallel to 6880 kHz.

5750 OM/SS dictates 5L message in phonetics to second OM/SS in USB at 1128. Phonetics "Carlos," "Golfo," "Julieta," "Miguel," "Noviembre," "Romero," and "Uniforme" used as substitutes for EE phonetics. Rest of SS phonetics same as in EE. Heard on Wednesday. • YL/GG in USB heard at 0300 with repeated 3F group and 1-0 count, ten dashes at 0305 and "gruppen," count, into 3-2F groups.

5762 YL/SS with 5F groups in AM 0400 and 0600. Active!

5770 YL with call-up of "Delta Tango" from 2130-2135 followed by 5F GG messages for "991," "503," "015," and "686." These addressees were all used by stations DFC37/DFD21 (East German stations) before they shut down in early 1993. Addressees now greatly reduced in number and reallocated to various GG two letter stations.

5771 YL/SS in AM with 5F groups 0814.

5800 OM/EE fishing vessels in USB 0325; X-rated topics.

5807 OM/EE with Russian accent repeating "261" from 2100-2105 followed by "945 945 38 38" and into 5F groups. Ended with "00000." Repeated next day on 5805 kHz at same time.

5812 YL/SS in AM with 4F groups at 0415.

5820 YL/EE in AM with 5F groups 0425.

5821 CW station sending "NNN," etc. from 2000-2030 followed by YL/GG with "gruppe 30."
• FEMA USB operations.

5841 Scrambled voice communications in USB at 0738.

5875 Strong carrier here at 2200. At 2204 OM/RR said "236" once. Then at 2215 OM/RR repeated "236 236 236," "000," and off at 2220.

5900 YL/SS in AM with 5F groups 1106.

5930 YL/SS repeating "996 123" in USB 0200.
• YL/SS in AM with 4F groups 0305; sometimes parallel to 10665 kHz.
• YL/SS in USB with 4F groups 2300 and 0000; sometimes parallel to 10665 kHz.

5961 FEMA USB operations.

6049 FEMA USB operations.

6106 FEMA USB operations.

6108 FEMA USB operations.

6151 FEMA USB operations.

6176 FEMA USB operations.

6227 Two OM/EE, possible drug smugglers, discussing choice of weapons and other preparations for Friday pick-up off Freeport. USB at 0207.

6228 YL/SS in AM with 5F groups 0800.

6242 Six dots, one fast dash, two dots, one dash rasper 1439.

6271.5 "DE IR IR QTC" repeated in CW 1400.

6293 YL/SS in AM with 5F groups 0900; badly distorted audio.

6425 Unidentified CW station sending "146" at 1703, then into 5F groups.

6500 YL with 5L phonetic groups in USB mode at 1539.

6507 YL/GG with 5F groups at 0527.
• "Swedish Rhapsody" tune in AM at 2100. After music box tune, broadcast programming came on. Later stopped and replaced by YL/GG with 5F groups.
• Strong AM carrier at 2345. At 2349 letter "U" repeated in CW until 2359. Then repeated "LOLO LOLO LOLO/01723" until 0005, then "BT BT" and into 5F groups. At end of message repeated "LOLO" call and message. Signed down with "AR SK AR SK."

6604 OM/SS outbanders in LSB on New York/Gander Volmet frequency at 0310.

6644 CW SS traffic network around 0130; calls include "65Z," "Y2L," and "Z3F." Plaintext traffic; appears to concern the police or military forces of some Latin American nation.

6660 White noise here at 2312. Had wide bandwidth of 6660 to 6720 kHz.

6675 CW station in AM mode sending "54768" from 2120-2125 followed by 5F groups.
• Outbander net in LSB 2230, apparently located in Britain and Ireland.

6681 Outbander 300 baud packet network, apparently in Europe; topics include propagation, satellite communications, antennas, and other communications subjects.

6683 Air Force One in USB, mainly working Andrews AFB.

6700 YL/SS repeating "924" and 1-0 count, then into 5F groups at 0300. This sked was on a Friday. The Wednesday schedule at 0200 had YL/SS repeating "35403."

6708 CW station at 1300 sending "EU6" (526) in cut number format. Then at 1305, three long tones, "526 526" and into 5F groups. This is usually a YL/GG transmission.

6714 YL in Chinese broadcast in AM at 1431. Sounded like "liao feng zhe hu zaio" (repeated three times), "o shi" (twice), "fe liao fend" (three times). Ended with single "fe gi" or "bai gi" at 1435. Almost identical broadcast previously heard on 10694 kHz and similar broadcasts heard on 5828 and 8860 kHz.

6720 "I3C," unidentified with long alphanumeric message. Parallel to 11255 and 18009 kHz.

6732 Anti-drug smuggling EE traffic in USB 0400; several references to El Salvador.

6735.1 "01977 64769 89631 11158 BT" repeated in CW 0042, then "X" continuously.

6736 Unidentified OM SS traffic in SSB 0441 interspersed with EE military-sounding traffic apparently about some sort of surveillance operation.

6745 YL in USB mode at 2047 repeats "Charlie India Oscar Two". Bad QRM from Chinese broadcast station. Also heard "Charlie India Oscar Two" call-up at 1847.
- "Victor Lima Bravo Two" repeated by YL in AM 0050.

6758 YL in Czech with repeated 3F groups, then into 5F groups. USB at 0230 on Friday, 0500 Sunday, and 2200 Thursday and Friday.

6768 YL/SS in AM with 5F groups 0700.

6778 YL/SS in AM with 5F groups 0600.

6780 CW dashes sent at 0030 followed by CW dot dashes sent at 0040. At 0042 "726 726 726 98 98 44 44" and into CW 5F groups. Zero was cut as letter T.

6784 YL/EE with 1-0 count and "009" from 2100-2110. After ten tones "count 198" and into 3-2F groups. Also on 5413 kHz.
• YL/SS in AM with 5F groups 0200.

6785 YL/SS repeating "atención 381 02" and into 5F groups in AM mode at 0200 on Saturday. Very strong signal.

6786 "WAM TW" call-up in CW 0203 followed by "ARM TW" and 5L groups; off at 0217 with "AR AR AR SK SK SK."

6790 SLHFM "V" heard at 0203.

6792 MCW 4F groups 0030; most messages only two to six groups in length.

6797 YL/SS in AM with 5F groups at 0507.

6800 YL/SS in AM with 4F groups 0314.

6801.5 SLHFM "S" heard at 1600.

6802 YL/SS in USB with 4F groups 2300.

6808 Sound resembling music played backward and at low speed heard from 0210 to 0430 over several days.

6809 FEMA USB operations.

6810 OM/EE fishing vessel network in LSB 0118 with profanity; apparently lobstermen off the northeast coast.

6812 Air Force One communications in USB.

6814 White noise jamming noted here at 0107.

6825 YL/SS in AM with 5F groups 0711.
• YL/GG in USB with 5F groups 0500.

6832 "861" repeated in CW 0602, followed by "395 395 64 64" at 0604 and into 5F groups until 0611.

6833 YL/GG at 0545 with 5F groups ending with "00000 ende."

6836 5F CW groups 2045.

6840 YL/SS in AM with 4F groups 0230 daily; also reported at 0312. **Active!**
• YL/SS in USB with 4F groups 0230; rarely heard.
• OM/SS in USB repeating "Repito, repito, grupo 221" 0230; caused interference to 4F numbers transmission.
• OM/SS voice traffic in USB 0300-0800; seems to be from Central America.

6851 Approximately 45 seconds of CW dashes, then "541 541 541 154TT 154TT 154TT" in CW followed by 30 seconds of CW dots at 0244. Sequence repeated and then off.

6853 YL/GG repeating "Romeo Delta" from 1900-1905 followed by 5F groups.
• YL/GG repeating "Echo Golf " with musical tones, then into 3-2F groups in USB at 2104 on Wednesday.

6854.2 5L MCW groups 0800.

6855.9 YL/SS in AM with 5F groups 0411; badly distorted audio.

6867 YL/SS in AM with 5F groups 0800.

6872 YL/SS in AM with 5F groups 0700.

6880 "Lincolnshire Poacher" station. YL/EE repeated "40136" from 1900-1910. Then into 200 5F groups. Jammed here and also on parallel 5746 kHz. Also heard at 2200 on Tuesday, Friday, and Saturday. YL has British accent and delivers numbers in slow, lilting cadence.

6888 YL/SS in AM with 5F groups 0500.

6890 Two-second loop of 18 tones in USB at 0500 repeating continuously; sounded like two bird calls spliced together.
• YL/SS in AM with 5F groups 0500 and 0700.

6892 YL/SS in AM with "atención 873 09 07" repeated from 0500 to 0505, then into 5F groups.

6920 YL/SS in AM with 5F groups at 0530.

6929.5 KKN50, U.S. Deptarment of State in CW at 0313 with hand-sent "VVV VVV KKN50 AR" and then back to marker tape. Active!
• Call signs of stations heard calling/working KKN50 on this frequency include KRC81, KWT91, and KWT94.

6933 YL/SS in AM with 5F groups at 0500 Sunday.
• YL/EE in AM with 3-2F groups 0907.

6934 YL/SS in AM with 4F groups at 0302 and 0426.

6935 YL/SS in AM with 5F groups 0500.
• YL/SS in AM with 4F groups 0410.
• YL/SS in USB repeating "atención 822 09," then into 5F groups at 0300 on Thursdays.
• YL/EE in USB with 3-2F groups 0000.

6937 One dot, one dash rasper 1432.

6942 YL/SS in AM with 5F groups 0600.

6945 Three dots, one dash rasper 1315.

6950 YL/GG in AM repeating "038 038 038 429 92 027 429 92 927" and into 3-2F groups on Saturday at 0102. This is the YL who says "noyner" for nine.

6953 YL/SS in AM repeating "02 100" at 0200 and into 5F groups at 0205. Heard on Saturdays.

6962 YL/EE in AM with 3-2F groups 0215.

6966 5L groups in CW 0236.

6968 White noise jamming here at 0120.

6995 OM/SS in USB at 0545, seems to be some sort of military traffic net with headers like "telegrama numero cero ocho" and mentions of "en punto commandante." Probably Mexican due to mentions of "Baja California del Sur" and other Mexican states.
• YL/SS in AM with 5F groups at 0200.

7030 Warbling tone 0330.

7039.4 SLHFM "F" heard at 1209.

7100 OM/EE in AM with 3-2F groups at 0400.

7348 FEMA USB operations.

7351 "TB6" in CW with hand-keyed "VVV" marker 0715.

7371 "378 4454 75" in CW 2130 followed by 5F groups.

7394.5 SLHFM "V" heard at 1600, 1700 and 2000.

7395 "Water gurgling" sound 0400.

7401 YL repeating "Kilo Romeo" with electronic tones from 1430-35. Then 5F GG groups for "156" totalling 86 groups in length.
• YL/GG in AM with 3-2F groups 0000.

7407.2 OM/RR traffic net in USB 2200.

7409 CW network around 1100 to 2130 with calls "GCA," "WHY, "AVO," "IGK," and "INU" as net control. Hand-sent CW; believed to be Cuban Border Guards.

7410 "FMQ" working "CML" with hand-sent CW 2345.

7415 YL/EE in AM with 5F groups sent twice 0510.
• "E" repeated in CW 0345. On one night, frequency was interferred with by an unlicensed "pirate" radio station, and the "E" was replaced by "FU" until the pirate left the air!

7416 OM/SS network in USB around 0000-1000; several references to cities in Panama and Colombia and content is related to business and operational matters of apparently legitimate shipping companies.

7417 "White noise" centered on this frequency, extending 2.5 kHz on either side.

7418.7 OM/EE network of clam diggers (apparently off Delaware coast) in LSB 0400.

7422 YL/SS in AM with 4F groups 0200; parallel to 11533 kHz.
• YL/SS in USB with 4F groups 0000 and 0300.
• YL/EE in USB with 4F groups 2300.

7423 CW station at 0530 sending "941 941 941 TTT." T's are probably cut zeros.
• YL/SS in AM with 4F groups 0402.

7425 YL/SS in AM with 5F groups at 0209.
• YL/SS in AM with 4F groups at 0200 and 0425.
• YL repeats "Kilo Papa Alpha Two" in USB at 2017.
• YL/EE repeating "Echo Zulu India Two" at 2200, 7 days a week.

7435 YL/SS in AM with 5F groups 0330.

7442 YL/SS with 4F groups in USB 0000.

7445 At 2100, call-up of "550" and 1-0 count, then into 5F groups. Repeated next three days on same frequencies/ times, parallel to 9090 kHz.

- Call-up of "Kilo Papa Alpha Two" heard in USB at 1816 and 2045.

7450 YL/EE repeating "4161 9528 5505" at 0030.

7452 SLHFM "R" in CW heard here at various times between 1200-0200.

7469 5F CW groups 0900, zero cut as "T." Automatic sending at high speed.

7470 Fishing vessels in USB 0600 with OM operators and X-rated conversation.

7470.5 YL/EE in AM at 1400 with "581 581 581" and 1 to 0 count, ten tones, "count 225" and into 5F groups.

7472 "741 741 741 1" repeated in CW 0545.

7473 One CW pulse per second 0520, synchronized precisely with WWV.

7475 Pulse-like data bursts lasting ten to twenty seconds per burst 0400-0500.

7480 OM/RR at 1800 with "362" in AM until 1805. Then "759 759 48 48" and into 5F groups. Ended with "00000."

- YL/SS in AM at 0115 with 5F groups.
- 5F CW groups 0445.

7482 YL/SS in AM with 5F groups at 0100 and 0900.

7495 "Warbler" tone 0800.

7522 5L groups in MCW 0800.

7525 YL/SS in AM with 5F groups at 0700.

7527 YL/GG in AM with 5F groups 1900.

7532 YL repeating "Mike Delta" from 2000-2005. Then message for "241 40, 296 72," "attention," and into 5F EE text. Station sending nine dots and one dash in background.

7540 YL passing 5L phonetic groups in USB at 1539.

7541 At 0030 ten tones followed by YL/EE repeating "5787 4563" and off at 0040.
• At 1015 YL/Serbo-Croat (Bulgarian Betty) sending 5F groups. Off with "konec konec" and CW "AAAA."
• YL/EE after ten tones at 0030 with "5836 7840 0559" parallel to 5205 kHz.
• YL/EE in AM repeating "8699 3003 2565" 0220
• SLHFM "V" at 2200.

7542.6 Illegal fishing communications from Florida area in USB at 0331.

7546 "Warbler" tone 0800.

7550 5L CW groups 0037, parallel to 5358 kHz; bad hum.

7558.6 "YAQ" working "YAO" with 5L groups and some SS text in 75 baud ASCII at 2350.

7588 At 1800 YL/EE with 1-0 count and "918" until 1810 when ten tones were sent. Then into 3-2F groups after Count 92.

7605 **Active!** Carrier on at 2040 with Mossad station "Juliett Sierra Romeo" traffic faintly in background. At 2045 "Victor Lima Bravo Two" started up. Proper frequency for "Juliet Sierra Romeo" traffic was 5091 kHz. At 1800, YL repeating "Victor Lima Bravo Fifty" and at 1850 "Victor Lima Bravo Sixty." Also on 4665 KHz.
• "Mike India Whisky Two" repeated by YL in AM 2016.

7626 CW station sending "QRA QRA QRA DE KWS78" at 1900, U.S. embassy station listed as being Athens, Greece.

7648 YL/SS in AM with 5F groups 0230.

7650 On Wednesday at 2000,. YL/EE repeating "636 636 636 80688 031" until 2007. After five dashes, went into 5F groups. YL said "I say again" and then repeated message.

7655 YL/SS in AM with "449 449 449" andd 1-0 count and into 3-2F groups 2134.
- YL/SS in USB with 4F groups 2100.
- YL/EE with 3-2F groups in AM 2115.
- YL/EE with 3-2F groups in USB 2100.

7662 KWL90, U.S. embassy (reportedly Manila, Philippines), with QRA marker in CW 1237.

7692 Seventeen dots, one dash rasper 1225.
- Nine dots, one dash rasper 1751.
- Five dots, one dash rasper 1244.
- Ten dots, one dash rasper 1916.
- Seven dots, one dash rasper 1519.

7695 YL/GG in AM with 3-2F groups 0323.

7698 OM/EE fishing vessels in USB 0630; mainly business topics.

7724 KRH50, U.S. embassy (reportedly London, England), with QRA marker in CW 0954.

7740 YL/GG repeating "Delta Tango" with electronic tones from 2130-2135, then into 5F groups for "991" and "503."

7755 YL/SS in AM with 5F groups 0200.

7760 5F groups in CW 1204, then "BT BT BT" and into steady tone.

7762 YL/EE in AM with 4F groups at 0136.

7763 At 0130, ten tones followed by YL/EE repeating "6358 5187 6209 3363" and off at 0140.

7780 Weird, haunting signal like "whale songs" or muted feedback from several microphones 2322.

7786 5F CW groups 0045, very sloppy hand-sent Morse.

7845 YL/SS in AM with 5F groups 0407; overmodulated and distorted. **Active!**

7858 YL repeating "Oscar Alfa" from 2200-2205 followed by 5F message in GG for addressee "820."

7860 YL/SS in AM at 0332 with 5F groups.

7864 OM/SS network in USB around 0630; tight net discipline and seems military or paramilitary in nature.

7871 YL/EE in AM mode at 1401 repeats "267 267 267" and 1-0 count until 1410, then 10 beeps, "count 225, count 225" and into 3-2F groups. YL/EE 3-2F broadcast can be found here on Sunday at 1400. YL/EE in AM mode at 1401 repeats "172 172 172" and 1 to 0 count. Noted with 3-2F groups at 1411.

7886.6 YL/SS in AM at 0304 with 5F groups.

7887 CW station with 5L groups using cut numbers ANDUWRIGMT. Went down with "AR AR AR SK SK SK" at 1017.
- CW station sending "M Q E Q" over and over at 0305. Ended with "AR."
- Lincolnshire Poacher tune in USB at 1506 alternating with YL/EE repeating "19134" until 1510, then three pairs of different tones and into 5F groups, each sent twice; was parallel to 8464 and 9251 kHz. Also heard 0544-0547 and 1708.
- YL/EE in AM with 5F groups 0230 and 0400.

• YL/EE in USB with 5F groups 0412; English accent.
• YL/EE in LSB with 5F groups 0500.

7888 YL/SS with 5F groups in AM 0300 and 0700.

7918 YL in USB at 1732 repeats "Yankee Hotel Foxtrot Two." Mossad.
• YL/EE in USB with 5L groups in phonetic alaphabet 0510.

8004 OM/EE on fishing boat talking to YL/EE ashore (apparently his wife) in LSB 0245. Conversation dealt with family matters.

8018 YL/SS in AM with 5F groups at 0600.

8026 YL/Serbo-Croat (Bulgarian Betty) sending 5F groups at 1100. Gradually building up traffic after going off air in late 1991. Can also be heard daily at 1400 on 5311 kHz.

8063 YL repeating "Delta Mike" with tones from 1900-05. At 1905, YL/EE with "message for 214, 20 groups," "attention" and into 5F groups.
• YL/GG repeated "Oscar Alpha" from 2200-2205 followed by 5F groups message for "820."

8074 OM/EE at 2015 with "947" call-up. Then "310 310 2 2 11111 93597 310 310 2 2 00000."

8075 YL/SS with 4F groups in AM 0430.

8080 YL/EE with 3-2F groups in AM 2312.

8120 YL/GG with 1-0 count and '791' from 2000-2010. After ten tones "gruppe 224" and into 4F groups. Also on 10135 kHz. This scheduled is every Saturday.

8127 YL/EE with "Charlie India Oscar Two" at 2348; off at 2351.

8135 At 2100, very strong carrier followed by continuous tone at 2105. Then at 2110 CW station sending "139 139 139" and into rapid CW 5F groups.

8136 YL/SS in AM mode repeating "atención 753 00" then into 5F groups at 0300 on Wednesday.

8142 CW traffic in 5F groups and YL/Czech at 0000, 0200, 0300, 0400, 0500, and 0600 preceded by "VVV DE OLX" in CW. Also heard on 5301 kHz. Other monitors heard schedules at 0900, 1100, 1300, 1800, 1900, 2000, and 2100.

8150 OM/SS in AM repeating "atención 456 00" and into 5F groups 0505.

8160 YL/EE with repeated "00000" and "574 574 574" from 0500-0510 in USB on Sunday, similar to OM/RR on 10125 kHz.

• YL/EE with 5F groups in USB at 0500 Saturday and Sunday, same YL as in Lincolnshire Poacher transmissions but that tune not used here. Numbers begin promptly on the hour.

8165 YL/SS repeating "atención 381 00" and into 5F groups in AM mode at 0101 on Saturday. Message repeated on 6953 kHz at 0201.

8173 YL/EE with "Delta Mike" from 1900-1905. Then "message for 214, 214 11 groups and 15 groups," "attention," then into 5F groups. YL said the same word for "and" and "end." Same message sent next day at 0900 on 17340 kHz and at 1900 on 9325 kHz.

• YL/GG repeating "Golf Kilo" from 1930-35 with electronic tones.

• YL repeating "Mike Delta" from 2000-2005. Then "message for 241/45 groups, 296/49 groups, attention" and into 5F groups in EE.

8186 YL/SS in AM with 5F groups at 0600, 0800, and 1100. **Active!**

8188 Music box playing "Swedish Rhapsody" melody every Sunday at 0800, 1000 and 1300. Then YL repeats 5F group for two minutes and then into 5F groups for that addressee.

8205 Screechy noises mixed with tones moving around in pitch. Could be voice scrambler because some apparent speech pattern was heard in all the din. USB at 2028.

8240 YL/SS in AM with 5F groups 0400.

8256 Cut number message in CW at 0237 using AU34567DNT = 1 to 0. Messages "890" at 0240 and "891" at 0242. Signed off at 0246.

8270 YL/EE in AM repeating "479" at 0501. At 0505, "678 678 678 42 42" and into 5F groups, each sent twice.

8300 Chinese numbers station "New Star Radio Station Number 4" signing on, then into YL/CC 4F groups, each sent twice, AM mode at 1103. Similar, but non-parallel broadcast noted on 11430 kHz in AM at 1108. "New Star" YL/CC in AM at 1235 with 4F groups. "New Star" station also heard in AM at 0203 with 4F groups each sent twice, parallel to 12750 kHz.

8312 YL/GG in AM with 3-2F groups 0221; parallel to 10135 kHz.

8380 YL/SS in AM with 5F groups at 0200.

8494.8 SLHFM "S" heard at 2329.
- SLHFM "C" heard at 2328.

8495.2 SLHFM "F" heard at 2241.

8559 At 1730 ten tones followed by YL/EE repeating "6582/7925/5551/2208" until 1740. Also on 5205 kHz.

8641 YL in USB mode at 1814 with 5L phonetic groups. Announced "end of message, end of transmission" at 1822, then repeated "Mike India Whisky" in phonetics til 1825 followed by "Message, message, group 58, group 58, text, text," then repeated same 5L phonetic message. Went through same routine at least twice more. Two faint, non-parallel 5L broadcasts heard in background. Another sked to "Mike India Whisky" heard at 2046.

8464 Electronic musical tones to the tune of "Lincolnshire Poacher" heard at 1505 in USB mode. Alternating with repeats of "966 66" until 1511 then four tones and into 5F groups with each group sent twice. Was parallel to 7887 and 9251 kHz. 8464 kHz was the weakest with heavy CW interference. On another day a schedule noted in USB at 2045. Warbling type jammer trying unsuccessfully to block the numbers transmission.

8465 YL heard repeating "Sierra Yankee November Two" in phonetics on various days at 1432, 1533, and 1732. All USB mode.

• YL heard in USB with 5L groups at 1537 followed by "end of transmission" within the minute. This was not parallel with similar transmission on 7540 kHz, which continued after 8465 kHz went down.

8494 SLHFMs "C" and "S" heard at 1525.

8495.2 SLHFM "F" heard at 1203.

8645.5 SLHFM "S" heard at 1600.

8698 SLHFM "B" heard at 0430; "buzzy" sounding.

8728 YL/SS in USB 0810 working other stations; sounds military in nature.

8752.1 OM in Oriental language 0830 in USB. OM was almost screaming and there were some notes played on a keyboard between messages.

8780 YL/SS in AM with 4F groups at 0107.

8805 YL/GG with very slow 5F groups. Ended with "378 378 22 22 00000" at 2015.

8840 Three clandestine broadcast stations and warble jammers playing cat-and-mouse games on this frequent and 8820, 8850, and 8870 kHz. Broadcast stations would change frequencies but jammers would follow.

8873 YL/SS in AM at 0914 with 5F groups. Went down at 0917.

8879 OM in unidentified language in USB 0252 apparently singing! Has also been heard around this time on 8915 kHz.

8889 OM/EE fishing vessels in USB 0600.

8906 QTH radar bursts at 0200 causing QRM to air traffic control communications.

8970 YL/SS with 5L groups in AM 0300.

8980 YL/SS in AM at 0512 with 5F groups. Very strong signal.

8989 OM/EE in USB with 5L groups using the phonetic alphabet. 0440. Once heard doubling with another transmission; on another occasion, was transmitted live and announcer laughed in middle of list.

9012.6 9 dots, 1 very long raspy dash heard at 1810.

9033.2 "Water gurgling" sound 0832.

9040 YL/GG repeating "Papa Delta" from 0430-0435, then 5F groups for "551" and "054." Another "Papa Delta" broadcast heard 1930 with 5F GG messages for "551" and "301."
• YL/EE (German accent) with repeated "Mike Delta" and musical tones followed by 3-2F groups in USB at 2000 Sundays. This is the same YL and format as used in the YL/GG "two-letter" transmissions.

9049 YL/EE repeating "352 252 352" and 1-0 count in USB 2106.

9065.5 KNY82, unknown U.S. government station, leaving the air after working WGY906, FEMA, Denton, TX, at 1745 in USB.

9090 YL/EE in AM at 2100 with 3-2F groups; sometimes parallel to 9755 kHz.
• YL/EE in USB at 2100 with 3-2F groups.

9092 Sound like a "walking man" in USB 0715.

9094 YL/EE in USB with 3-2F groups 2100.

9120 WGY912, FEMA Special Facility, Mt. Weather, VA in CW at 1519 with three characters per group (alphanumeric characters).
• YL/SS in AM with 5F groups 0600.

9130 YL/EE in USB mode at 1902 repeats "Echo Zulu India," then "two messages, message, message, group 26, group 26, text, text" and into 5L groups. Strong signal. Also heard on another morning at 1734 with similar 5L traffic.

9140 YL/SS in AM with 5F groups 0206.

9222 YL/SS in AM with "928 928 928" and 1 to 0 count 0303.
• YL/SS in AM with 4F groups 0300.

9237 YL/SS in AM with 5F groups 0500 and 1100.

9240 Foghorn signal at 0345.

9250 OM/RR with "124" from 1300-05, then into 5F groups but stopped after eight groups. Then started up with "124" again until 1311 when into 5F groups ending with "249, 249, 38, 38" and "00000."
• YL/EE in AM with 5F groups 2220, ended with flute-like tones.

9251 OTH radar bursts here at 0421. Also noted on 11233, 11857, and 12764 kHz at various times.
• YL/EE with 5F groups in USB 0618; English accent.
• "Lincolnshire Poacher" station with YL/EE giving 5F groups in AM 2103.
• YL/SS in USB with 5F groups 0415.

9255 YL/SS in AM with 5F groups 0400.

9274 YL/EE in AM at 1502 repeats 1-0 count, and "125, 125, 125" until 1510, then 10 beeps, "count 225, count 225" and into 3-2F groups. Message repeated at 1531, and was parallel to 11123 kHz. This transmission seems to be scheduled Sunday and Wednesday s here at 1500. Another night heard same operator at 1 627 with untypical 4F groups.

9325 YL repeating "Mike Delta" with electronic tones from 1930 to 1935. Then "message for 241, 251, 60 groups, attention" and into 5F groups in EE.

9328 OM/RR at 2055 with 5F groups. Ended with "000 000."

9330 YL/SS in AM at 0304 with "atención 256 03" repeated. At 0305 "03 110" repeated. 0306 into 5F groups.

9381 5F groups in CW at 0318. Repeated groups and ended with "BT BT 530 530 128 128 TTTTT."

9394 "L9CC" with 5F CW groups 0224.

9450 YL repeating "Golf Kilo" from 1930 to 1935, then 5F GG message for "4TI."

9462 FEMA USB operations.

9570 Warbling jammer at 2000 jamming unidentified broadcast station.

9649 YL/EE in AM mode at 1305 repeats "249 249 249" and 1-0 counts til 1310, then ten beeps, "count 225, count 225" and into 3-2F groups parallel to 12168 kHz. Was getting stepped on badly by Adventist World Radio Korean Service broadcasting on 9650 kHz.

9651 CW station at 2135 sent message at approximately 15 wpm. "WRA U6G-CC-NS-DN-TR-IMI." Repeated few more times and ended w/AR at 2139.

9725 YL/CC with 4F groups each sent twice parallel to 11430 kHz in AM mode at 0727. Similar activity noted on the hour during evening on 8300, 11430, 15388, and 13750 kHz. YL/CC passing 5F groups, each sent twice, AM mode at 1450. YL/CC 4F sent twice commonly noted here, so 5F was somewhat unusual.

9755 YL/EE in AM with 3-2F groups at 2100; parallel to 9090 kHz.

9785 YL/SS with 4F groups 0415 in USB.

9831 YL/EE in AM repeating "131 131 131" and 1 to 0 count parallel to 12221 kHz,

9833 5F CW message with unusual characters at 1200.

9843 YL/EE in AM with 3-2F groups at 2245.

9958 YL/SS with 4F groups in AM 0230.

10060 OM/SS traffic in USB 1300; sounded military in nature and mentioned several locations in the Yucatan area of Mexico.

10125 YL with call-up of "Charlie India Oscar Two" in USB at 1349 to 1350. Also heard skeds at 1845, 2045, and 2348.
• OM/RR repeating "00000" and "835 835 835" from 0511 to 0515 Sunday.
• YL/SS in AM with 5F groups at 0330.

10161 "Y7G" with 5F CW groups 0042.

10162 OM/EE with "947" from 1920-1925 then "183 183 207 207" and into 5F groups. Male version of YL transmision where she pronounces number 8 as "ATE." Heard previous evening at same time but with different message.

10177 YL with "November Zulu" from 2200-2205, then 5F GG groups for "649" and "955." Repeated a week later on 9450 kHz.
• YL/GG repeating "Charlie November" with musical tones and into 3-2F groups in USB at 0300 on Thursdays.
• YL repeating "Whiskey Lima" from 1300-1305, then 5F groups for "522" of 88 groups in length.

10179 OM/RR in AM mode with 5F groups at 2106 Saturdays.

10180 YL/SS in AM with 5F groups 0300.

10194 FEMA USB operations.

10235 YL/GG with "465" from 2115-2120. Then "972 972 15 15" and into 5F groups. Down with "00000."

10243 OM/EE repeating "241 241 241 000" from 2020-2025.

10250 OM/EE repeating "192" at 2000 until 2005 when sent "860 860 37 37" and into 5F groups. Ended with "00000."

10262 YL/EE in AM mode at 1321 passing 3-2F groups, parallel to 13906 kHz. Heard on Saturdays at 1400. Another day heard in USB at 1305 with 3-2F groups.

10270 YL/SS in AM with 5F groups 0400.

10284 "VVV DE 8BY" in CW at 0648 followed by trinome groups.

10285 SLHFM "V" heard at 0300 and 1500.

10345 YL/SS in AM with 5F groups at 0500.

10345.5 Unidentified CW station sent "NUG TM" until 0403, then "AGM TM" til 0405 and into 5L groups until 0418. Ended with "AR AR AR SK SK. " This cut number system used letters T, M, I, A, N, D, G, U, R, and W.

10374 At 1800 YL/GG with 1-0 count and "853." At 1810 after ten tones, "gruppe 186" and into 3-2F groups.

10410 Station "LN2A" in CW w/unusual RTTY modes in between at 1908.

10424 Strong carrier at 1740 with tune-up tones until 1800. Then YL/GG with "620 620 620 00000" and off at 1805.

10464.2 "QRA DE KUN50 QSX 10/12/14/17/24 K." Same CW speed and marker format as KKN50. Heard at 0357.

10493 WGY912, FEMA Special Facility, Mt. Weather, VA with daily roll call of stations, including WGY904 (District 4, Atlanta), WGY905 (District 5, Chicago), WGY906 (District 6, Denton, TX), WGY907 (District 8,

Kansas City), WGY908 (District 8 Denver), WGY909 (District 9, San Francisco) WGY910 (District 10 Bothell, WA), WGY911 (Washington, DC telecommunciations headquarters), WGY916, 918, 950 (all unidentified) and WGY901 (District 2, Boston) called, QSY F-27. Heard at 1651. At 2341, WGY908 working Play Ball (an aircraft) with phone patch to "7525." Play Ball does radio check and breaks patch. All in USB. On another day, WGY912 working WGY910, 906, and 902 with radio checks. Advises no national test today due to shortage of staff. Heard at 1803. At 1824, White House Communications Agency working WGY912 with radio check. At 1959, "Whiskey Hotel" base (White House) with WGY912 for radio check. These communications also in USB.

10510 YL/SS in AM with 5F groups 0400; sounded as if it were taped and running a little too fast.

10529 YL/EE in USB at 1317 with 3-2F groups. On at same time as 10262 kHz transmission but not parallel to it. YL/EE in AM mode at 1344 passing 3-2F groups. parallel to 16198 kHz. YL/EE in AM at 1348 with 3-2F groups. Non-parallel transmission noted on 11626 kHz.

10533 YL/SS in AM with 4F groups 0415; sometimes parallel to 7422 kHz.

10546.5 YL/SS in AM with 4F groups 0325.

10588 FEMA USB operations.

10597 YL/EE in AM mode at 1424 passing 3-2F groups, parallel to 13609 kHz. Can be found on these two frequencies Friday nights at 1400.

10610.2 OM/SS in USB at 0241 dictates messages with 3L groups in SS to second OM/SS.

10611.9 SLHFM "C" in CW 2150.

10643.5 SLHFM "S" heard at 1600.

10648 YL passing 5L phonetic groups in USB mode at 1347. Also heard at 1837 with similar traffic. YL/EE in AM at 1346 with 5L groups followed by "end of message, repeat, repeat message. Group 96. Group 96. Group 96" and into a repeat of the 5L groups message.

10665 YL/SS in AM at 0303 with call-up "967 967 967" and 1 to 0 count.
• YL/SS with 4F groups in AM 0000.

10694 YL/CC in AM at 1204 repeats something like "yiao du liao fu jiao" three times, "o shi" twice, and "ba sent liao" three times until 1207, then single "bai gi," and carrier abruptly down. Heard something very similar on 5828 kHz several weeks before, though that was followed by what sounded like one side of a telephone conversation.

10702.3 Unidentified CW station at 2008 with 5L groups with pause after every ten groups.

10713 YL/SS in AM with 5F groups 0530.

10723 YL/EE in AM at 1406 repeats "267 267 267" and 1 to 0 counts til 1410, then ten beeps, "count 225, count 225" and into 3-2F groups.

10740 OM/RR repeating "567 567 567 00000" from 2130 to 2135, then off.

10766 OM/EE fishing vessel net in LSB 0249.

10820 OM/EE in AM mode at 1104 repeats "135" until 1104, then into 5F groups, each sent twice. Mechanical-sounding male voice with odd accent, probably computer generated. Ended with "00000" at 1112.

• Unusual Mossad activity for few days. At 1740, YL repeating "Mike India Whisky 14B05B88." Next day at 1938, "Mike India Whisky 14D37 D38 D39 D40 D41 D42 D43 D44 D45 D46 D47." At 1914 on another day, "Mike India Whisky 14D76D77D78." At 1700, "Mike India Whisky 14B07" and one hour later, "Mike India Whisky 14B88." Also on 8641 and 5230 kHz.

10860 OM/EE repeating "431" from 2115-2120. Then "367 367 125 125" and into 5F groups. Ended with "00000."

10865 YL/SS in AM with 5F groups at 0730.

10868.3 WGY912, FEMA Special Facility, Mt. Weather, VA in CW with following message: "DE WGY912 KS7 BOL DGY AAD KDS Q5X X76 B5N WOC A6N 2YU 13I DKX 5DG MUE JHP BAM." Heard at 1546. Message repeated every minute on the minute. Listened to several reports and dropped at 1549.

10870 FEMA CW operations.

10871 SLHFM "D," Odessa in CW at 1907. SLHFM "C," Moscow in CW at 1907.

10871.9 SLHFM "S," Arkhangelsk, Russia in CW at 1756. Just above this was SLHFM "C," Moscow, Russia which was heard at 1756 also.

10872.7 SLHFM "F" heard at 1155. Distinct 50 Hz hum on the signal which was not present on any of the other "F" markers.

10940 YL/EE in AM mode at 1132 with F groups. This station noted infrequently with 4F groups, though normally passes 3-2F groups.

10970 YL in USB heard this frequency with "Victor Lima Bravo Two" call-ups at skeds of 1345, 1447, 1756, 1846, and 2050. YL also heard with call-up of "Victor Lima Bravo India Twelve" at 2010 which was also on 7605 and 4665 kHz.

11000 CW station at 1055 sending "VVV DE OLX." At 1100, YL/Czech sending ?29; first digit not audible due to fault on tape or voice device. At 1105 into 5F text but number 7 (sedm) was not working properly. Deduced original 3F group was "729." Parallel broadcast on 6760 kHz was OK.

11062 "555 817 817 817 23" in CW 1900, into 5F groups 1905. Full digits were sent instead of cut numbers.

11090 YL/GG in AM repeating "823 823 823" and 1 to 0 count at 2310, followed by ten tones and into 3-2F groups.

11107 YL/GG with 3-2F groups in AM 0013.

11108 YL repeating "Mike Delta" from 2000-2005 followed by "message for 241 70 groups, attention." Then into EE 5F groups. In background, could hear CW station sending "360 360 360 00000." This frequency was previously used solely by "Papa November" til it ceased operation. Reportedly, the transmissions ending with three or five zeroes (or cut zeroes) are SVR (ex-KGB) communications to agents.

11120 OM/RR with 5F groups, each sent twice, AM mode at 1408.

11123 YL/EE with 4F groups in AM at 1822. Was same operator usually heard with 3-2F traffic, though 4F texts heard several time before and since on this frequency especially.

11125 YL/SS in USB at 0205 repeats "atención 521 08," no 1-0 count. At 0209, repeats "08 32." At 0210 into 5F groups with no beeps preceding text. "Final, final" at 0212 and then repeats from beginning.

11129.9 OM/SS network in USB around 2130 to 2320. Believed Mexican army; tight net discipline and mention of various Mexican cities and states.

11130 Scrambled voice transmissions in USB 2030.

11139 YL/GG at 2110 repeating "917 917 917 000" and off at 2115.

11146 YL/EE with 5F groups in USB at 0139.

11160 YL/EE in AM repeating "425" at 0124, then "693 693 122 122" and into 5F groups sent twice.

11173 OM/Italian in USB switching to RTTY 0200, seems to be Italian diplomatic communications channel due to message content.

11174 OM/EE net in USB at 1836; apparently U.S. petroleum engineers under contract to Pemex (the Mexican state oil company) from content of messages.

11188 11 dots and 1 dash signal sent at 2112. Raspy signal.

11200 OM/EE in AM at 1534 repeats "135 135 135" and 1-0 count. Ten beeps at 1539, "count 61, count 61" and into 3-2F groups. Heard on Saturdays.

11226 Air Force One communications in USB. **Active!**

11260 OM/EE (RR accent) at 2005 with "897" call-up. (Very strong carrier.) Followed by "106 106 49 49" and into 5F groups. Ended with "00000." Repeated at 2100 on 8660 kHz.

11415 YL/EE in AM at 1538 with 3-2F groups parallel to 13418 kHz.

11430 YL/CC in AM at 1231 with 4F groups, each sent twice. Very strong signals. This is similar to a non-parallel transmissions on 8300 kHz which are on at same time. This probably the "New Star Radio Station #4" CC numbers station.

11440 At 1800 YL/GG sending "461," then "406 406 128 128" and into 5F groups. This YL had same voice as an East German "four organ note" station which ceased operation in 1990.

11455.7 "QRA DE KKN50," U.S. Department of State, Washington DC, with QSX marker in CW at 0347.

11460 YL/SS in AM with repeated "699 699 699" and 1-0 count, then into 4F groups at 0300 Fridays.

11468 YL/SS in AM with 5F groups at 0200.

11470 YL/EE in AM at 1209 repeats "617 617 617" and 1 to 0 counts til 1210, then ten beeps, "count 225, count 225," and into 3-2F groups. Message repeated at 1231, then down at 1251. Parallel to 16198 kHz. "Warble" jammer noted here on another day at 1246.

11467 Strong DSB carrier here at 1920 followed by extremely rapid CW sending of "733," nearly too fast to read. Then into faster yet CW, too fast to even recognize as being CW.

11479 Unidentified CW station with traffic of 5F groups. Uses letter T as cut number for zero. Heard at 1623.

11485 EE/YL in AM with 5F groups 0815.

11491 YL/SS in USB with 4F groups at 0323 on Sunday.

11532 YL/SS in AM with repeated "699 699 699" and 1 to 0 count then into 4F groups at 0300 Fri.day

11570 YL/EE repeating "528" in AM at 1703, then "190 53" and into 5F groups. Signed off with "190 190 190 53 53 00000."

11615 Air Force One channel for LSB communications.

11626 YL/EE with 3-2F groups in AM at 1320, paralle to 14408 kHz. There were also concurrent, but non-parallel, transmissions on 16198 and 10529 kHz.

11628 "2X" and "2A" passing number codes at 0124 in USB.

11640 YL/EE with 5F groups in AM 0715; has Irish accent!

11721 FEMA USB operations.

11801 FEMA USB operations.

11957 FEMA USB operations.

12009 FEMA USB operations.

12084.5 YL/SS repeating "655" and 1-0 count 2200, ten tones, "grupo 115" and into 4F groups.

12092 YL/GG two-letter station near "Oscar Alpha" here at 1500 and "Juliet Whiskey" at 2100. BBC World Service nearby on 12095 kHz.

12135 OM/RR in AM at 1402 with 5F groups, each sent twice.

12143.7 YL/SS in AM with 5F groups 1307.

12155 YL/SS in AM with 4F groups 0320.

12166 YL/EE with "507" call-up and 1 to 0 count, into 3-2F groups 1505.

12167.8 YL/EE in USB with 3-2F groups 1505.

12190 OM/RR in AM at 1404 with 5F groups, each sent twice.

12203 OM/EE at 1920 repeating "821" until 1925, when he said "697 697 40 40" and into 5F groups. Repeated next day on 12210 kHz at same time. Very strong signal.

12207 OM/EE with 3-2F groups in AM 1956.

12210 OM/EE with "821" from 2020-2025 then "759 759 34 34" and into 5F groups. Down with "00000." • KWA80, U.S. embassy at an unknown location, QRA marker in CW 1530. Suspected locations include Tokyo and Seoul.

12212 CW station repeating "791 791 791 27 833 833 833 24 555 555 555" then sent 27 group message for "791" and 24 group message for "833." Signed off with "000."

12214.2 CW station at 1201 with cut numbers. "RIW TR TR AGM BT" repeated three times and into text.

12216 WGY912, FEMA Special Facility, Mt. Weather, VA, with WGY905 (Chicago) and WGY908 (Denver) in USB 1716 for encrypted data test.

12221 YL/EE in AM at 1205 repeated "409 409 409" and 1 to 0 count til 1210, then ten tones, "count 225, count 225" and into 3-2F groups parallel to 13555 kHz.

12225 OM/RR with "169 169 169 000" from 2010-2015.

12242 YL/EE with 1 to 0 count and "187" from 2100-2110. After ten tones, "count 184" and into 3-2F groups.

12282 "8BY" with "VVV" marker in CW 1956.

12283 "DEA47" with VVV marker in CW 1457.

12314 YL repeating "Mike Delta" from 1930-35. Then "message for 241, 241, 69 groups, attention" and into EE 5F groups.

12319 YL/EE in AM mode at 1506 repeats "255 255 255" and 1-0 counts til 1510, then ten beeps, "count 225, count 225" and into 3-2F groups parallel to 14703 kHz.

12356 Two OMs in Greek in USB 0147, with apparent argument via radio.

12420 At 1900 very strong carrier tuning on this frequency until 1930 when OM/RR sent "618" for 5 minutes. At 1935, "537 537 2 2 11111 11111 00055 00055 537 537 2 2 00000" and then off. This message consisted of two groups only (11111 and 00055).

12455 CW station at 2109 sending 5F groups. Cut zero as "T," rest all sent full.

12603 YL/EE "Lincolnshire Poacher" station at 1509 with "29040" repeated in between tones. Also on 11545 and 13375 kHz.

12618 YL/EE in AM at 1502 repeats "981 981 981" and 1-0 counts parallel to 14423 kHz.

12747 YL in USB mode at 1433 repeats "Mike India Whisky Two" in phonetics.

12950 YL in USB at 1327 repeats "Kilo Papa Alpha Two" in phonetics.

13257 One dot, one dash rasper 1234.
- Nine dots, one dash rasper 1702.
- Three dots, pause, 14 dots rasper 1914,

13313.4 Unidentified CW station at 1755 sending message using cut number system AU34567DNT = 1 to 0.

13345.6 OM/EE fishing vessel net in USB 1344 with X-rated language.

13355.5 Unidentified CW station sends "QTT6 ZJQK" and then calls "SYNC DE FWGZ QTC K," repeats call-up and into message heading "1811 256 BT ZH F14T BT" and into 5F groups. Sends "K" upon completion but no response heard on this frequency. Operator was using a speed key. Heard at 1528.

13365 OM/EE in AM at 0734 repeats "286 286 286 1," then "652 67, 652 67" and into 67 groups of 5F traffic. Passed text only once, then down with "000, 000." Voice sounded like it was computer generated but with a heavy accent.

13375 YL/EE with 5F groups in USB 1615.

13380 "TW MW BT" repeated in CW 1930 and into 5L groups.

13385 KKN39, U.S. Department of State, Washington, DC, QRA marker in CW around 0000.
• Call signs of stations heard calling/working KKN39 on this frequency include KWB69, KFB90, KEA32, KMA52, KGA66, and KLM49.

13389.6 OM/SS network in USB 1319. Believed to be Venezuelan military due to tight discipline, military titles (colonel, etc.) used, and references to Venezuelan cities.

13404 5L CW groups 2339.

13410 5L CW groups 2116, each group sent twice. Then heard "BC8 BC8 NC8 BC BC DE. . . OK QRU QRX WA 0000Z. . . EE INT C QRX WA 0000Z SURFX QJT BND C AR"

13415 OM/RR at 1815 repeating "154" then "906 906 32 32" and into 5F groups. Ended with "00000."

13416 YL/EE in AM repeating "021 021 021" with 1 to 0 count at 1203, ten tones at 1210, "count 210, count 210," and into 3-2F groups.

13419.5 Unidentified CW station at 1611 with cut numbers. Ends with "AR AR AR SK SK SK." Off at 1618.

13420 OM/EE said "821" once at 1906. Later at 1920 OM/EE repeated "821" until 1925, then "604 604 33 33" and into 5F groups.

13423 YL/EE in AM with repeated 3F group and 1-0 count from 1800-1810, then into 3-2F groups. Heard Tuesday and Saturday parallel to 16174 kHz.

13446 "Rich Lady," USAF tactical callsign working FEMA WGY908, Denver, CO in USB at 1607.

13447 KKN39, U.S. Department of State, Washington, DC (listed site) with QRA marker 2305 and numerous transmitter spurs on 13271, 13327, 13502, etc.

13450 YL/SS in AM with 4F groups 0000.

13452 YL/SS with 4L groups in AM 1315; direction-finding techniques placed this signal as originating near Remington, VA.

13471 KGA64, location unknown, calling KKN50, U.S. Department of State, in CW 1747. KKN50 responds and says to QSY to 17660 kHz.

13487.3 Low tone followed by higher and then burst of 6-7 seconds. Heard from 1603-1612. Similar transmission but very faint heard in background. Pauses between transmissions varied in duration.

13518 YL/EE in USB at 1733 with 3-2F groups.

13533 Israeli Mossad station YL/EE at 1814 in USB with 5L groups in phonetics.

13544 "963 963 973 1" sent slowly in CW 1925, then "362 124" sent fast and into 5F groups with zero cut as "T."

13555 YL in AM at 1209, repeats "934 934 934" and 1-0 count until 1210. Then ten beeps, "count 225, count 225" and into 3-2F groups.

13555.9 CW traffic addressed to "Direccion Salud Galapagos" 2045, believed to be Ecuadoran Navy.

13560 OM/SS communications in LSB 2115; informal network with poor discipline (whistling, sounds effects, etc.).

13565 YL/SS with 5F groups in USB 0214.

13567 Data bursts 2128.

13600.1 Unidentified CW station at 1445 sending two-, three-, and four-character groups in very long mesage; believed to be Vietnamese diplomatic traffic.

13635.5 HFSLMs "S" heard at 1600, "D," "P," "C," and "S" at 1000, "D" and "C" at 1400, "D," "C," and "S" at 1800, "D," "P," "S," and "C" at 0900, "D" and "C" at 2000, "C" at 0300, "D" and "C" at 1500, "D," "P," and "C" at 1000, "P," "C," and S" at 1700.

13636.5 SLHFM "F" heard at 1158.

13636.8 SLHFM "S" heard at 0340.

13652 Unidentified CW station at 1316 sending cipher text in continuous strings of characters. Some Cyrillic characters noted.

13668 One dot, one dash rasper 1702.

13775 YL/GG repeating "Whiskey Lima" from 1900-1905 then 5F groups for "026" and "522".
•YL repeating "Juliet Whiskey" from 2100-2105, then 5F GG groups for "542" and "824."

13795 YL/SS in AM with "526 526" and 1-0 count, "grupo 83," and into 4F groups at 1305.

13815 KRH50, U.S. embassy (reportedly London, England), in CW 0448 with QSX marker.

13890 YL/GG repeating "Echo Golf" from 0900-0905 and then into 5F groups for "823." Message was 100 groups in length.
• YL/GG repeating "Whiskey Lima" from 1930-1935 followed by 5F groups for "522" and "026."
• OM/RR with "289" from 2000-2005 followed by "863 863 59 59" and into 5F groups. Ended with "00000."
• YL/Yiddish in USB 0711 with 5F groups, each sent twice.

13906 YL/EE in USB at 1437 w/3-2F groups.

13921 YL in USB mode at 1448 repeats "Charlie India Oscar Two" in phonetics, unusually strong signal. Also heard this call-up on another day in USB from 1549-1550. YL in USB at 1434 with call-up "Victor Lima Bravo Two."

14180 YL/SS in USB with 5F groups 0410.

14330 OM/EE in USB at 2106 with 5F groups sent twice.

14352 "H4W" hand-sent repeatedly 1603.

14360 "QRA DE KWS78," U.S. Embassy, Athens, Greece with QSX marker in CW at 2114.

14363.8 Hand-sent CW using cut numbers 1330; sender would pause after each five groups and stop after each 50 groups to allow receiving station to request repeats.

14392 YL/EE in AM with "804" call-up, ten tones, and into 225 3-2F groups 1510, parallel to 15450 kHz.

14406 YL/EE with 3-2F groups in AM mode at 1347.

14408 YL/EE passing 3-2F groups in AM at 1328 parallel to 11626 kHz. Concurrent, non-parallel transmissions on 10529 and 16198 kHz.

14421.1 5L groups in hand-sent CW 1425.

14423 YL/EE with 3-2F groups in AM at 1504. Also heard at 1514 on another day with this broadcast parallel to 14703 kHz, and at 1549 on yet another day.

14435 "451 451 451 TTT" repeated in CW 2042.

14442 OM/EE in AM mode at 1431 passing 5F groups. Sounded like computer generated voice.
• OM/EE (with foreign accent) in USB at 1512 repeated "645 645 645 000" until 1517.

14445 5F CW groups 1450, automatic sending with zero cut as "T."
• "555 555 817 817 45" repeated in CW 1900.

14450 FEMA USB operations.

14481.8 5L groups in "fluttering" CW 1609, zero sent as "T."

14487 YL/EE with 5F groups in USB at 1320.
• At 1400 on same day noted "Vive le Companie" and "906 28" call-up by YL/EE and into 3-2F groups at 1410. Also with 5F groups at 1426 on another day.

14488.6 Warble jammer 1303.

14494.5 OM/EE repeating "120 120 120" and 1 to 0 counts in USB 1607. Ten dashes at 1610, "count 55," and into 3-2F groups.

14546 YL/EE in AM at 1421 passing 3-2F groups.

14564 YL/EE in AM mode at 1408 repeats "456 456 456" and 1-0 counts. Noted with 3-2F groups at 1411.

14620 OM/EE in AM repeating "620 620 620 00000" at 1904.

14622.8 Unidentified station in FSK Morse at 1555 sending what sounds like "ETFNJX TKAGL" over and over, then off.

14646.5 Packet "bulletin board" system, participants include military contractors (Harris Corp., etc.) and U.S. Army stations.

14648 Packet "bulletin board" system, apparently operated by U.S. Air Force; most text is EE but some SS noted along with call signs of USAF stations, such as AFA01 in Washington, DC.

14655 "QRA DE X4X" in CW 2237, off abruptly at 2239.

14662 Foghorn heard at 2233.

14675 OM/RR repeating "231 231 231 00000" from 0900-0915, then off.

14687.7 5L CW groups 1420, sounds hand-sent; ends with "BT AR."

14703 YL/EE in USB at 1522 with 3-2F groups parallel to 14423 kHz.
• YL/EE in AM at 1624 with 3-2F groups. Heard on many days and at various times between 1400 and 1700.

14710 YL/EE in USB at 1403 repeats "Echo Zulu India." At 1404 begins 5L groups in phonetics.

14727.6 YL/GG with 3+2F groups each group sent twice, USB at 1741.

14736 Unidentified CW station with cut number traffic at 1408.

14749.5 YL/EE repeats "Mike India Whisky Five Six Zulu Two" 1439 followed by data transmission.

14750 Israeli Mossad call-up "Charlie India Oscar Two" in USB at 1949.

14752.2 YL/EE repeats "Charlie India Oscar Two" in USB at 2043 for approximately 10 seconds. Then second station, YL/EE, heard giving 5L groups in phonetics. First YL returns 2044 with "Charlie India Oscar Two" call-up till 2049.

14770 Unidentified CW station (sounds like CLP1, Havana) sending cut numbers at 1322.

14776 WGY912, Mt. Weather, VA FEMA Special Facility heard at 1729. This frequency is FEMA channel 36. Working WGY909, San Francisco, CA at 1732 for message from WWJ63 (SHARES Nashville, TN) to NMC (SHARES, San Francisco) re future SHARES exercise and damage assessment engineer. Switched to F26 (10493 kHz) at 1736.

14815 CLP65, Cuban embassy, Managua, Nicaragua calling CLP1, MFA, Havana, Cuba in CW at 1503.

14825 Cut number groups, automatic CW, at 1456 ending with "AR AR AR SK SK SK."
- YL/SS in AM with 5F groups 1308.

14837 FEMA USB operations.

14886 FEMA CW operations.

14892 Nine-tone sequence in AM 0622.

14899 FEMA USB operations.

14908 FEMA USB operations.

14950.3 Unidentified automatic CW at 1829 with 5F groups. Down at 1831 with "000 000."

14977 OLX, PTT, Prague, Czech Republic in CW at 1257 with "VVV" marker. At 1300 YL/Czech repeats "725" followed by 5F groups from 1303-1310. Next day heard again at 1257 with call-up followed by 5F groups. Zero was cut as letter "T" in CW. This station also noted at 0855 with "VVV DE OLX" marker for five minutes followed by YL/Czech repeating "177" til 0905, then into 5F groups. Also on 8142 kHz.

14983 SLHFM "E" heard at 1400.

15460 YL/EE in AM with 3-2F groups 1500.

15481 YL/EE in AM with "333" call-up 1310, directly into 3-2F groups without any tones 1310.

15651 YL/SS in USB with 4F groups 1230.

15675 OM/SS in USB 1330; noise like a siren at end.

15682 Electronic tones in USB at 1401 with Lincolnshire Poacher tune alternating with repeats of "62898" by YL/EE. At 1410, three double-tones, and into 5F groups, each sent twice. On another day, noted heavy jamming on top of numbers station.

15704 Pips, one every two seconds. Heard at 1503 on ISB. Pips speeded up to one per second at 1510 and slowed back down to one every two seconds at 1513.

15938 YL/SS in USB with 3-2F groups at 1300.

15970 KKN50, U.S. Dept. of State, Washington DC with QRA/QSX marker at 2234.

16083.9 YL/EE in USB with 5F groups 1341; "warble" jammer atop.

16084 YL/EE repeats "59687" and "Vive le Compagnie" tune. At 1610, hi-lo tones one octave apart followed by 5F groups. Also heard at 1734 on another day.

16086 YL/EE in AM at 1207 with 3-2F groups. Also heard at 1645 on another day.

16198 YL/EE in AM with 3-2F groups from 1345 to 1351. Next day same YL/EE on at 1300 with "323 323 323" and 1-0 count, repeated. YL/EE with 3-2F groups heard another day this frequency and parallel to 10529 kHz at 1348.

16201 FEMA USB operations.

16226 Unidentified station with "TTT" (cut number zero) followed by burst transmission, then "TTT" and another burst. Heard at 1341.

• "261 403 126 BT BT" in CW 2305 followed by 5F groups, each sent twice.

16241.1 Unidentified CW station (suspected Vietnamese diplomatic) at 2218 with two-, three-, and four-letter groups.

16245.3 Station using speed key CW at 2102 sending 5F groups using AU34567DNT = 1 to 0 cut number system.

16280 OM/RR with 5F groups, each sent twice. Heard in AM at 1308.

16306 YL/SS in AM with 4F groups at 1830.

16317 YL/EE repeating "652" in AM 1700; at 1739, "809 809 154 152" and into 5F groups sent twice.

16326.9 "674 674 674 TTT" repeated in CW at 2120.

16341 CLP8, Cuban embassy, Conakry, Guinea in CW at 1258 calling CLP75 and requesting QSV. No reply heard this frequency.

16414 YL/GG repeating "Kilo Whiskey" from 0930-0935 and then into 5F groups for addressees "091" and "023." YL repeating "Kilo Whiskey" from 1430-1435 followed by GG 5F groups for "513."

16430 FEMA USB operations.

16434 At 1300, long tone followed by carrier until 1330, when ten tones were sent. Then YL/EE repeating "8273 8276 4187" til 1340; also on 13423 kHz. Another day this daily sked noted with 10 dashes followed by 4F groups with two or three groups per message. On this occasion YL/EE repeating "3975 2846 0622" over and over til 1340. Also on 13423 kHz and both frequencies troubled by "warble" jammers.

16459 YL/EE in USB with 4F groups at 1421.

16652.1 Seems to be private "chit chat" network for maritime radio operators in CW. Many references to fish, docking, ports, etc. Most communications in EE with lots of jargon. Calls heard include "JVR," "5B," "AA," "PARE," "RQ," "LAN3," "LAN2," "Pat," and others. Heard around 1600.

17016 SLHFM "C," Moscow heard in CW at 1320 and also heard another day at 1330.
• SLHFM "S" in CW 2349.

17410 Possible Mossad transmission noted here in USB mode at 1709. YL passed two 5L phonetic messages (group counts 6 and 59), with repeats before announcing "end of messages, end of transmission" at 1721.

17425 YL/SS in AM with 5F groups 0030.

17427.3 MFA, Havana to Cuban embassy, Lima, Peru at 1753 in RTTY 75/500 with 5F groups.

17437.1 Unidentified station at 1751 with hand-sent CW sending cipher text (some cut numbers, some full numbers) which were sent in a continuous string of characters. Message heading was message, number, group count, and BT. Ended with "BT AR." Cut number systemsused was AU34567DNT = 1 to 0.

17479.8 "A2 DE D1 Y42 AAA MSG NR 90834 BK WMF ZPS VAH OGX RZJ K" repeated in CW 2214, followed by trigraphic groups.

17485 YL/EE in USB with "601 601 601" and 1 to 0 count (repeated) at 2105.
• YL/EE in AM 0134 with 4F groups.

17519 FEMA USB operations.

17520 YL/EE win AM with 5F groups 1950.

17560 YL/GG in AM 1837 with 3-2F groups.

17649 FEMA USB operations.

17966 YL in USB mode at 1449 repeats "Charlie India Oscar Two" in phonetics. Very slowly and deliberately sent.

17970 YL with Asian accent repeating "Charlie India Oscar Two" from 0550 to 0551. Another Mossad station.

18035 YL/SS in AM at 1303 with 5F groups, very weak signal.
• YL/SS in AM repeating "atención 950 01" then into 5F groups at 2300 on Monday

18164 CLP1, MFA, Havana, Cuba working unidentified station in CW at 1759.

18173 5L CW groups 1225, ends with "VVV" at 1234.

18195 YL/GG with 3-2F groups in AM at 1612.

18200 MCW cut numbers 1503, strong signal slopping from 18185 to 18227 kHz.

18237.4 4-note musical marker in USB at 2346 with carillon-type sound. Marker ended at 0002 and then a fluttering sound began.

18434 Unidentified CW station at 1540 sending 5L groups using cut number system ANDUWRIGMT. Went down with "AR AR AR SK SK SK" at 1554.

18456.9 CLP1, MFA, Havana, Cuba working CLP8, Cuban embassy, Conakry in RTTY 50/500 at 1506 with news items in FF. Message was a "CIRCULAR" message addressed to Cuban embassies in French speaking countries.

18486.2 CW station calls "759" for approximately ten minutes, nothing further and carrier goes off. Heard 1600-1610.

18565.5 16-note musical marker heard in USB from 1341-1345.

18628.2 CLP1, MFA, Havana, Cuba working Cuban embassy, Guyana, in RTTY 75/500 at 1429 with RYs and into 5F groups.

18630 CLP1, MFA, Havana, Cuba in CW at 1456 working unidentified station, sending traffic in SS.

18707 YL/EE in USB with 3-2F groups at 1825.

18744 FEMA CW operations.

18843.6 Two OM/EE exchanging signal reports in USB 1627. Weaker station tells other this frequency is "18 Bravo" and they should go to 22159 kHz, which is "22 Alpha." Weaker station says he is glad he can be heard and will continue checking for signal reports all the way up to San Diego. Stronger signal call is WUZ6854.

18881 YL/EE at 1600 Saturday/Sunday calls "509" and then into 5F groups each sent twice. On another day, message number sent at 1700 was repeat of message previously sent on 22222 kHz at 1600.

18887 YL/SS in AM with 4F groups at 1630.

18994.9 CLP5, Cuban embassy, Moscow, "DE CLP1," MFA Havana in CW at 1740 requesting QSV. Other end not heard.

19095 YL/EE in USB with 3-2F groups at 1501.

19113 440 Hz tone in USB at 1458 heard every two seconds; 1459 every second with double tones every 10 seconds. 1501 every second but 440 Hz tone on odd seconds and 880 Hz tone on even seconds. 1504 back to 440 Hz tone every two seconds and sequence repeats. Heard during periodic checks throughout the day.

19162 Unidentified CW station apparently repeating group requests. After last report sends "VVV QSL" and signs off.

19180 CLP1, MFA, Havana, Cuba with news items in SS. RTTY 50/500 at 2104. Items dealt with speech by Fidel Castro.

19183.2 RTTY 75/500 at 2101 with news items for Cuban embassies sent by CLP1, MFA, Havana, Cuba.

19333 CLP1 to unidentified Cuban embassy with SS message at 1835 in CW.

19490 Unidentified CW station at 1819 with 5F groups, zero cut as letter T. Followed by messages in SS. Went to RTTY with encryption 1825-1829, returned to CW at 1830 with more 5F group traffic. Down at 1837.

19757 FEMA USB operations.

19802 CLP1, MFA, Havana, Cuba with RTTY traffic (50/500) at 1542 to EMBACUBA Guinea, Bissau and Conakry.

19954 Russian manned spacecraft telemetry and beacons; beeping sounds interrupted at intervals by "trilling" and "burring" sounds.

19969 FEMA USB operations.

20008 Russian Soyuz manned space vehicle communications, both AM and CW.

20027 FEMA USB operations.

20096 Six tone marker, notes C-E-C-E_b-C-D, in USB 1300.

20115 YL/EE in AM with 3-2F groups 0020.

20152 Unidentified CW station at 1532 sending "QRA A1L II CC II PM II PB II VZ II IMI" repeated til 1534.

20220 YL/EE in AM with 3-2F groups at 2141.

20250 YL/EE in AM with 5F groups at 2110, each sent twice.

20316 YL/SS in AM with 5F groups at 1830.

20350 YL/GG repeating "Charlie Delta" from 1100-1105 followed by a 47 group message of 5F groups for "707."
• 1600-1605 YL repeating "Hotel Kilo" then YL/GG with 5F groups for "393" and "621."
• YL/GG in AM with 3-2F groups 1606.

20365 YL/EE in USB with 5F groups 0335; strong Australian accent.

20368 YL/EE in AM at 0326 with 3-2F groups, parallel to 22970 kHz.

20405 YL/EE in AM 1800 with "758 758 758" and 1-0 count, followed by 3-2F groups at 1810.

20448.6 CLP1, MFA, Havana, Cuba with RTTY transmission 50/500 at 1819 with news In SS for Cuban embassies.\

20587 "JQN DE BXZ" and "KGJ DE BXZ" in hand-sent CW 1726; "BXZ" sends 5F groups, zero cut as "T."

20818.9 CLP12, unknown Cuban embassy told to QSV by CLP1, MFA, Havana, Cuba. CW at 1558.

20821 Foghorn blasting several times at 2155. CLP1, MFA, Havana, Cuba in CW at 1555 sending "VVV" strings.

20832.8 CLP1, MFA, Havana, Cuba with 5F traffic for EMBACUBA Congo. RTTY 50/500 at 2048. Also heard 75/500 RTTY here on another day.

20853.4 CLP1, Havana. Must be working a relay station because he sent 5F groups for EMBACUBA Syria and then indicated also had traffic for Djakarta, Indonesia.

20929.4 Unidentified CW station at 1834 using a speed key for cut number traffic. System was AU34567DNT = 1 to 0.

20991.5 SLHFMs heard as follows: S at urday "D" at 1600, "P" at 1100, "D" and "C" at 1400, "D," "P," and "C" at 0900, "D" and "P" at 1500, "D" at 1000, "C" and "S" at 1700.

20992 SLHFM "S," Arkhangelsk and SLHFM "C" Moscow heard at 1530.

21830 OM/SS in USB at 2220. Located at Spanish embassy, Managua, Nicaragua and is talking to MFA, Madrid, Spain.

21865 YL/SS in AM with 5F groups at 0000 Sunday.

21956 Two-note musical marker in USB at 1750. Heard for very long period of time.

22110 YL/EE in AM with 5F groups at 1400, each sent twice; ended with "00000."

22222 YL/EE at 1600 on Saturday repeating "104" then "627 627 , 39 39" and into 5F groups, each sent twice. Ended with "00000."

23083.7 CLP1, MFA, Havana to EMBACUBA Nigeria. RTTY 50/500 at 1451 with 5F groups, then into plain text SS message to the Cuban ambassador in Nigeria.

23642 KWS78, U.S. embassy, Athens, Greece in CW at 1505 with "QRA DE KWS78 QSX 7/10/14/18/23 K."

23915 At 1550 in AM, two tones: lo-hi, short-long, 2 seconds cycle.

24467 Four-tone marker, lo-hi, lo-hi, every 2 seconds. USB at 1716.

24555 FEMA USB operations.

26010 CB outbanders in USB at 2152 discussing channels.

27095 Strong carrier came up at 1330. At 1500 YL/EE repeated 1-0 count and "512" call-up until 1310 when ten tones sent and YL went into 3-2F groups.

27527.6 French language traffic in SITOR-A 1440; appears bootleg.

27537.6 Bootleg packet radio bulletin board 1440; from items posted, seems to be in French Guiana.

27553 "DE ROBY FROM NORTH ITALY" in 45/200 RTTY around 1400.

27994.5 "LUCA" calling CQ in CW at 1907; probably an outbander.

29700 OM network in unidentified language in AM mode; apparently a "gypsy taxicab" disptaching system somewhere in the New York City area.

29715 EE telephone paging system in narrow-band FM; apparently somewhere in the West Indies and noted during periods of sporadic E propagation

29790 OM/SS network in USB 1700, apparently similar to U.S. outbanders.

29800 OM/RR communications in AM; noted during sporadic-E propagation and believed to be from Cuba.

29805 SS communications in AM with "whistling" tones between messages; believed to be mobile telephone communications from Mexico.

29845 SS communications in AM with "whistling" tones between messages; believed to be mobile telephone communications from Mexico.

29890 SS communications in AM with "whistling" tones between messages; believed to be mobile telephone communications from Mexico.

29950 OM/RR communications in AM; noted during sporadic-E propagation and believed to be from Cuba.

29980 U.S. outbander channel for narrow-band FM communications.

31268 YL/SS in AM with 5F groups at 1800; noted during sporadic-E propagation.

31685 YL/SS in AM with 5F groups at 1600; noted during sporadic-E propagation.

36150 YL/SS in AM with 5F groups at 1903; noted during sporadic-E propagation.

57455 "Radio American 306" in LSB working "LSB111" 2057. LSB111 gave location as "Apple Valley," while Radio American 306 said he would have "to quit due to getting in too deep."

APPENDIX A

References

Military Cryptanalytics, Part I, vols 1 and 2, Friedman and Callimahos.

Military Cryptanalytics, Part II, vols 1 and 2, Friedman and Callimahos.

Traffic Analysis and the Zendian Problem, Callimahos (This title contains material from *Military Cryptanalytics* Part II, vol 2, but with message texts in much larger print)
NOTE: All the above titles are reprints of declassified U.S. Government training publications.

The Defection of Igor Gouzenko, 3 vols, Report of the Royal Commission.
NOTE: All four of the above titles are published by Aegean Park Press, PO Box 2837, Laguna Hills, CA 92654.

The Code Breakers, David Kahn, The Macmillan Co.

Sword and Shield: Soviet Intelligence and Security Apparatus, Jeffery T. Richelson, Ballinger Publishing Co.

Soviet Signals Intelligence (SIGINT), Desmond Ball, Strategic and Defence Studies Centre, Research School of Pacific Studies, The Australian National University, Canberra, Australia.

U.S. Army Signals Intelligence in World War II, James L. Gilbert and John P. Finnegan, U.S. Government Printing Office.

Secret Warfare: The Battle of Codes and Cyphers, Bruce Norman, Distributed in the U.S. by Sterling Publishing Co.

*Widows, W*illiam R. Corson, Susan B. Trento, Joseph J. Trento, Crown Publishers.

KGB: The Inside Story, Christopher Andrew and Oleg Gordievsky, Harper Collins Publishers.

Every Spy A Prince: The Complete History of Israel's Intelligence Community, Dan Raviv and Yossi Melman, Houghton Mifflin Company.

Spy Catcher, Peter Wright, Viking Penguin Inc.

Spy vs. Spy, Ronald Kessler, Macmillan Publishing Co.

Tower of Secrets, Victor Sheymov , U.S. Naval Institute Press.

By Way of Deception: The Making and Unmaking of a MOSSAD Officer, Ostrovsky and Hoy, St. Martin's Press.

Inside the KGB, Kuzichkin, Pantheon Books.

Breaking with Moscow, Arkady N. Shevchenko, Alfred A. Knopf.

Secret Signals: The Euronumbers Mystery, Simon Mason, Tiare Publications.

Guide to Embassy and Espionage Communications, Tom Kneitel, CRB Research Books.

How to Tune the Secret Shortwave Spectrum, Harry Helms, TAB Books.

Shortwave Listening Guidebook, second edition, Harry Helms, HighText Publications, Inc.

"Single Letter HF Beacons," *SPEEDX Reference Guide to the Utilities.*

DOD Dictionary of Military and Associated Terms, JCS Pub 1, U.S. Government Printign Office.

Top Secret: A Clandestine Operator's Glossary of Terms, Bob Burton, Paladin Press.

Latest Intelligence, Tunnell, TAB Books.

This is the Federal Emergency Management Agency, Brochure L135, FEMA (P.O. Box 70274, Washington, DC 20024).

Dictionary of Electronics, Radio Shack.

Encyclopedia of Electronics, TAB Books.

Popular Communications magazine carries articles and columns containing information on Underground signals. Here is a short list of some pertinent articles:

"DXing the World's Embassies", Harry Helms, ***Popular Communications,*** September 1983.

"Monitoring MOSSAD: The Israeli Intelligence Service," Greg Mitchell, ***Popular Communications***, July 1984.

"High Frequency Single-Letter Beacons," William Orr, ***Popular Communications***, December 1984, January 1985 and February 1985.

"A New Family of HF Radio Beacons," William Orr, ***Popular Communications***, June 1986.

"Secrets of Shortwave Espionage," Don Schimmel, ***Popular Communications,*** May and June 1988.

"The Ghost of County Line Road," Havana Moon, ***Popular Communications***, October 1988.

"Those Mystery Signals," Don Schimmel, ***Popular Communications: 1993 Communications Guide.***

Monitoring Times magazine has similar articles, and there is also a specialty SWL club, The Association of Clandestine Radio Enthusiasts (ACE), which covers pirate, underground and clandestine radio activities in its monthly bulletin. For further information about club membership write to ACE, PO Box 11201, Shawnee Mission, KS, 66207-0201.

APPENDIX B

Prosigns and Abbreviations

SOME SINGLE AND COMBINED Morse characters form what are called "operating prosigns." These are used by CW operators to convey bits of information pertaining to various phases of communications operations.

Included in this listing are abbreviations commonly used by SWLs when they are preparing loggings. Certain ones may also be seen in use on the air in much the same manner as the prosigns.

AA	Unknown station, all after
AB	All before
AGN	Again
AR	End of transmission
AS	Short wait
B	More to follow
BK	Break
BN	Between
BT	Long break
C	Correct
CFM	Confirm
CHARAC	Character(s)
CK	Sometimes used in place of GR (group count)
CLD	Called
CLG	Calling
CQ	All stations callup
CS	Callsigns

CT	Net control system (NCS)
CUL	See you later (later sked)
CUT NBRS	Letters are sent for numbers
DE	From
DIP(L)	Diplomatic
DISSEM	Dissemination
DTOI	Date/time of intercept
DTG	Date/time of group
DX	Signals from great distance
EEEEEEEE	Error
EMB	Embassy
EMBACUBA	Cuban embassy
ETA	Estimated time of arrival
ETD	Estimated time of departure
FB	Fine business (good)
FH	First heard
F, FIG	Figure(s)
FM	From
FOLL	Followed, Following
FREQ	Frequency
FSK	Frequency shift keying
GA	Go ahead
GB	Good-by
GR	Group count
GRP	Group(s)
HDNG	Heading
HQ'	Headquarters
HR	Here
HRD	Heard
ID, IDENT	Identification
II	Separative sign
IMI	Repeat or ?
INFO	Information
INT	Interrogatory
INTEL	Intelligence
K	You may now transmit
L, LTRS	Letter(s)

LANG	Language
MFA	Ministry of Foreign Affairs
MIN	Minute(s)
MKR	Marker
MSG	Message
N	No, Negative
NCS	Net Control Station
NIL	I have nothing to send, no traffic
NR	Number
NW	Now
NX	News
O	Operational immediate precedence
OM	Old man
OPR	Operator
OPS	Operations
OS	Out station
P	Priority precedence
PBL	Preamble
POSS	Possible
P/P	Phone patch
PREV	Previous(ly)
PROB	Probably
PSE	Please
PT	Plain text
R	Roger, received
RCD	Received
RDO	Radio
RE	Regarding
REF	Reference
RPT	Repeat(s) (ed)
RQST	Request
RY's	RYRYRY (RTTY test)
SIG	Signature
SK	End of schedule
SKED	Schedule
SPEC	Special
SRI	Sorry

STA, STN	Station
SVC	Service Message
SWL	Shortwave listener
T	Station called to transmit to all or designated addressees. Also used as Zero in cut numbers systems.
TFC	Traffic
TKS	Thanks
TMW	Tomorrow
TO	Action Addressee(s)
TOI	Time of intercept
TU	Thank you
TXT	Text
UNID, U/I	Unidentified
UR	Your
UTE	Related to utility communications
VE	Understood. Also sometimes used in place of BK
VY	Very
W	Words count in message. Sometimes used in place of CK or GR.
WA	Word after
WB	Word before
WKD, WKG	Worked, working
WRKG	Working
WPM	Words per minute. Morse code speed
WX	Weather report
XCVR	Transceiver
XMT	Transmit
XMTR	Transmitter
XMSN	Transmission
X2, X3	Repeated two times, repeated three times, etc.
Y	Emergency precedence
YL	Young lady
Z	Flash precedence
5F, 5L	5-figure, 5-letter
73	Best regards

APPENDIX C

Glossary

additive A single digit, a numerical group, or a series of digits which, for the purpose of encipherment, is added to a numerical cipher unit, code group, or plain text, usually by cryptographic arithmetic.

addressee The component or individual to whom a message is directed by the originator.

alphanumeric Set of symbols consisting of letters and numbers.

AM Amplitude modulation, a mode of transmission where signal is transmitted by using the audio portion to vary the strength of the carrier wave portion.

American Black Chamber A group headed up by Herbert Yardley which did code/cipher breaking. Discontinued by Secretary of State Stimson, who reportedly said "Gentlemen do not read other gentlemen's mail."

anagram Plain language reconstructed from a transposition cipher by restoring the letters of the cipher text to their original order.

aperiodic system A system in which the method of keying does not bring about cyclical phenomena in the cryptographic text.

authentication A security system or measure designed for communications systems to prevent fraudulent transmissions.

autokey system An aperiodic substitution system in which the key, following the application of a previously arranged initial key, is generated from elements of the plain or cipher text of the message.

band A given set of radio frequencies.

base station Fixed land station communicating with one or more mobile stations.

baud Bits of data per second. (45 baud = 60 WPM; 50 baud = 67 WPM; 57 baud = 75 WPM; 75 baud = 100 WPM)

Baudot alphabet A five-unit code applied to teleprinter systems by Jean Maurice Emile Baudot. It employs a 32-element alphabet designed particularly for telecommunications wherein each symbol intended for transmission is represented by a unique arrangement of five mark or space impulses.

beacon Station which transmits signal that is used for navigational purposes by ships and aircraft. Station identification letters are usually sent continuously.

biliteral Cryptosystems, cipher alphabets, and frequency distributions which involve cipher units of two letters or characters (see digraphic).

bipartite alphabet A multiliteral alphabet in which the cipher units may be divided into two separate parts whose functions are clearly defined, e.g., row indicators and column indicators of a matrix.

blind Sending traffic without making contact with the recipient. Messages can be acknowledged at a later time and/or via other means.

book cipher A cipher system, utilizing any agreed-upon book, in which the cipher identifies a plain element present in the book.

break-in procedure The receiving station will break-in on the transmitting station by sending one or more long dashes or BK.

When the transmitting station stops, the receiving station sends the group number of the first letter of the group wanted repeated.

burst A short signal consisting of a number of cycles. The message is stored in memory or on tape and then transmitted at a very fast rate. This type of transmission is used to avoid interception and not be susceptible to direction finding. Also called *squirt transmission.*

bury To conceal or hide certain words, phrases or messages in the text of a communication.

callsign Any combination of letters, numbers, or a combination of both, used as the identification for a communications facility, command, authority, activity, or unit; used for establishing and maintaining communications.

carrier A signal which is modulated with the information to be transmitted.

channel A channel is a particular band of frequencies to be occupied by one signal, or one two-way conversation in a given mode.

chatter Exchange between operators of comments relating to traffic passed or personal interest items.

chi-square test A mathematical means for determining the relative likelihood that two distributions derive from the same source.

chi test A test applied to the distributions of the elements of two cipher texts either to determine whether they are the result of encipherment by identical cipher alphabets, or to determine whether the underlying ciphor alphabets are related. Also called the cross-product sum.

cipher component The sequence of a cipher alphabet containing the symbols which replace the plain text symbols in the process of substitution.

cipher system Any cryptosystem in which cryptographic treatment is applied to plain text units of regular length, usually monographic or digraphic.

cipher text The text of cryptogram which has been produced by means of a cipher system.

ciphony Enciphered telephony. Also called scrambling. The process of converting vocal communications into an unintelligible form and to change them back into an intelligible form through cryptographic treatment.

code book A book or document used in a code system, arranged in systematic form , containing units of plain text of varying length (letters, syllables, words, phrases or sentences) each accompanied by one or more arbitrary groups of symbols used as equivalents in messages.

codress A transmission or message with the address buried in the encrypted or encoded text.

coincidence test The kappa test, a statistical test applied to two ciphertext messages to determine whether or not they both involve encipherment by the same sequence of cipher alphabets.

columnar transposition A method of transposition in which the cipher text is obtained by inscribing the plain text into a matrix in any way except vertically and then transcribing the columns of the matrix.

communication The transfer of meaningful information from one location to another.

communication intelligence (COMINT) Information derived from the study of intercepted communications.

communication security (COMSEC) The protection resulting from all measures designed to deny to unauthorized persons information of value which might be derived from a study of communications. Crypto security and transmission security are the components of communication security.

compromise The availability of classified material to unauthorized persons through loss, theft, capture, recovery by salvage, defections of individuals, unauthorized viewing, or any other physical means.

concealment system A method of secret communication so designed as to convey a secret message without its presence being suspected by others than the addressee. In its most usual form, the plain text elements are concealed by combining them with extraneous plain text elements in such a way that the end result is an intelligible and apparently innocent message.

consumers Users of intelligence.

continuity Identify with respect to a series of changes. In cryptanalytic procedure, the maintenance of continuity involves keeping current a systematic record of changes in such variable elements as indicators, keys, discriminants, code books, etc., on a given cryptochannel. In traffic analysis, the maintenance of continuity involves the tracing of changes in call signs, frequencies, schedules or other variable elements assigned to a given radio station, link, or net.

control The station located at a headquarters or designated as being in charge of a particular radio net. Abbreviated as CT or NCS.

Coordinated Universal Time (UTC) Time scale based on the second, as defined and recommended by the CCIR and maintained by the Bureau International de l'Heure (BIH). For most practical purposes associated with the Radio Regulations, UTC is equivalent to mean solar time at the prime meridian (0 degrees longitude) formerly expressed as GMT. Many military services designate UTC as Z (Zulu) time.

covert A concealed or secret relationship.

crib 1. Plain text assumed or known to be present in a cryptogram.
2. Keys known or assumed to have been used in a cryptogram.

cryptanalysis The analysis of encrypted messages; the steps or processes involved in converting encrypted messages into plain text without initial knowledge of the key employed in the encryption.

cryptogram A communication in visible writing which conveys no intelligible meaning in any known language, or which conveys some meaning other than the real meaning.

cryptographic arithmetic The method of modular arithmetic used in cryptographic procedures which involves no carrying in addition and no borrowing in subtraction.

cryptography That branch of cryptology which treats of the means, methods, and apparatus for converting or transforming plain text messages into cryptograms, and for reconverting the cryptograms into their original plain text form by a simple reversal of the steps used in their transformation.

cryptolinguistics The study of those characteristics of languages which have some particular application in cryptology (e.g., frequency data, word patterns, unusual or impossible letter combinations, etc.).

cryptology That branch of knowledge which treats of hidden, disguised, or encrypted communications. It embraces all means and methods of producing communication intelligence and maintaining communication security; for example, cryptology includes cryptography, cryptanalytics, traffic analysis, interception, specialized linguistic processing, secret-inks, etc.

cryptosystem The associated items of crypto material and the methods and rules by which these items are used as a unit to provide a single means of encryption and decryption. A cryptosystem embraces the general cryptosystem and the specific keys essential to the employment of the general cryptosystem.

cut numbers Digits sent in abbreviated fashion. There are many different cut number systems in use throughout the world. They range from a simple system of just one number being cut as with zero sent as the letter T, to those with a letter substituted for each of the digits 1 to 0.

CW Continuous wave. Electromagnetic waves generated as a steady train of identical oscillations. They can be interrupted according to a code, or modulated in amplitude, frequency, or phase in order to convey information.

decipher To convert an enciphered message into its equivalent plain text by reversal of the cryptographic process used in the encipherment. (This does not include solution by cryptanalysis.)

decode To convert an encoded message into its plain text by means of a code book. (This does not include solution by cryptanalysis.)

decrypt See *decipher.* Also applies to decrypted, but not translated, message.

delta I.C. Index of coincidence applied to a small sample. See *index of coincidence.*

depth 1. The condition which results when two or more sequences of encrypted text have been correctly superimposed with reference to the keying thereof. Sequences so superimposed are said to be "in depth." 2. The number of such superimposed sequences, as "a depth of three."

diagnosis In cryptanalysis, a systematic examination of cryptograms with a view to discovering the general system underlying these cryptograms.

digraph A pair of letters.

dinome A pair of digits.

direct standard cipher alphabet A cipher alphabet in which both the plain and cipher components are the normal sequence, the two components being juxtaposed in any of the non-crashing placements (see *non-crashing*).

discriminant A group of symbols indicating the specific cryptosystem used in encrypting a given message. Also called the *system indicator.*

doublet A digraph or dinome in which a letter or a digit is repeated (e.g., LL, EE, 22, 66, etc.).

DSB Double sideband. Radio transmission in which both sidebands produced by modulation are transmitted equally along with a carrier.

double transposition A cryptosystem in which the characters of a first or primary transposition are subjected to a second transposition.

down at ____ Used by monitoring operator to indicate time station being copied ceased transmission.

dropped at ____ Used by monitoring operator to indicate coverage of a target transmission was discontinued at the indicated time.

dummy messages Bogus messages transmitted as part of a deception program to mislead enemy traffic and cryptanalysis efforts.

duplex operation Operating method in which transmission is possible simultaneously in both directions a telecommunication channel.

DXer Individual who listens to distance radio signals.

essential elements information (EEI) The critical items of information regarding the enemy and the environment that are required to make timely decisions.

electronic communication The use of electrical signals to send and receive information.

ELINT Electronic intelligence. Intercept of non-communications transmissions such as jammers and various types of radar.

emission The waves radiated into space by a transmitter.

encipher To convert a plain text message into unintelligible language or signals by means of a cipher system (encrypt).

encode To convert a plain text message into unintelligible language by means of a code book.

end product Finished intelligence.

factoring 1. An arithmetical process of determining the period of a periodic polyalphabetic cipher by a study of the intervals between repetitions. 2. In transposition, the process of determining column lengths by studying intervals between elements.

fading The characteristic of radio waves which causes their received amplitude to decrease. Fading is caused by a combination of several radio waves that are not in exact phase.

fixed A permanent installation.

flush depth 1. The condition which results when two or more encrypted messages have been encrypted starting at the same point in the key. 2. The number of such messages, as a flush depth of three.

frequency 1. In cryptology, the number of actual occurrences of a textual element within a given text. 2. In radio, symbolized by *f*. The number of recurrences of a periodic phenomenon in a unit of time. Radio frequencies are normally expressed in kHz at and below 30,000 kHz and in MHz above this frequency.

frequency distribution A tabulation of the frequency of occurrence of plain text, ciphertext, or codetext units in a message or a group of messages. A frequency count.

FSK Frequency shift keying. Modulating a radio carrier by changing its frequency by a small amount.

garble An error in transmission reception or encryption. Can cause word or phrase to be unreadable.

group A number of digits, letters, or characters forming a unit for transmission or for cryptographic treatment.

half duplex Transmission in both directions, but not simultaneously.

ham A radio amateur; slang for a licensed radio operator hobbyist.

heading Message externals preceding the text. May include serial numbers of various types, date-time group, precedence symbols, routing instructions, addresses, group count and possibly special instructions. The prosign BT is usually placed between the end of the heading and the beginning of the text.

HF High frequency, the portion of the spectrum from 3 to 30 MHz.

identification 1. In cryptanalysis, determination of the plain text value of a cipher element or code group. 2. In traffic analysis, determination of the specific unit, aircraft, ship, or Order of Battle involved in a given instance, but not its location.

index of coincidence The ratio of the observed number of coincidences expected in a sample of random text of the same size. Commonly known as I.C. (see also *delta I.C.*).

independent sideband Radio transmission in which each sideband corresponds to one or more modulating signals independent of modulating signals of the other sideband.

indicator In cryptography, an element inserted within the text of heading of a message which serves as a guide to the selection or derivation and application of the correct system and key for the prompt decryption of the message (see also *discriminant* and *message indicator*).

in here Used to indicate entry point of copying an intercepted item when the message is already in progress. Usually placed in brackets as (in here at (time)).

intelligence The product resulting from the collecting and processing of information concerning actual and potential situations and conditions relating to foreign activities and to foreign or enemy-held areas. The processing includes the evaluation and collation of the information obtained from all available sources, and the analysis, synthesis and interpretation thereof for subsequent presentation and dissemination.

intelligence community The members of the U.S. intelligence community are the Central Intelligence Agency, Bureau of Intelligence and Research of the Department of State, Air Force, Army, Navy, Marine Corps Intelligence elements, Federal Bureau of Investigation, Department of the Treasury, Department of Energy, the Department of Defense office that collects specialized national foreign intelligence through reconnaissance programs, and members of the intelligence community staff.

intelligence cycle Steps by which information is assembled, converted into intelligence and made available to consumes (users). Phases of the complete intelligence cycle are planning and direction, collection, processing, production and analysis, and dissemination.

intelligence reporting/evaluation system Method used to describe and rate the source of information as well as the content.

A	Completely reliable	1	Confirmed by other sources
B	Usually reliable	2	Probably true
C	Fairly reliable	3	Possibly true
D	Not usually reliable	4	Doubtfully true
E	Unreliable	5	Improbable
F	Reliability cannot be judged	6	Truth cannot be judged

intercept Message or messages obtained by interception. Also to engage in interception.

interception The process of gaining possession of communications intended for others without obtaining the consent of the addressees and ordinarily without delaying or preventing the transmission of the communications to those addressed.

interference The presence of unwanted signals or noise that increases the difficulty of radio reception. There are three kinds of interference: natural noise, manmade noise, and manmade signals.

interval A distance between two points or occurrences, especially between recurrent conditions or states. The number of units between a letter, digraph, code group, etc., and the recurrence of the same letter, digraph, code group, etc., counting either the first or second occurrence but not both. Frequently called *cryptanalyst's interval.*

ionosphere Outer part of the earth's atmosphere. The ionosphere consists of a series of constantly changing layers of ionized molecules. Many radio waves reflect back to earth from the ionosphere.

jamming Transmission set up to interfere with other programming on the same or a nearby frequency. Jamming methods vary and include varying tones, buzzing sounds, warbling, etc.

kappa plain constant A mathematical constant employed in coincidence tests such as the phi text, to denote the probability of a coincidence of a given plain text element or unit. It is the sum of the squares of the probabilities of occurrence for the different textual elements or units as they are employed in writing the text, for example, in English telegraphic plain text, the monographic and digraphic plain constants are .0667 and .0069 respectively.

key 1. In cryptography, a symbol or sequence of symbols applied to successive textual elements of a message t to control their encryption or decryption. 2. In radio, refers to a device used to send Morse code characters.

keyed columnar transposition A transposition system in which the columns of a matrix are taken off in the order determined by the specific key, which is often a derived numerical key.

keyword-mixed alphabet An alphabet constructed by writing a prearranged key word or key phrase (repeated letters, if present, being omitted after their first occurrence), and then completing the sequence from the unused letters of the alphabet in their normal sequence.

lambda test A test for monoalphabeticity in a message, based on a comparison of the observed number of blanks in its frequency distribution with the theoretically expected number of blanks both in (a) a normal plain text message of equal length and (b) a random assortment of an equal number of letters. Also called the *blank-expectation text.*

link The existence of direct communication facilities between two points.

logs Records kept by radio listeners of their shortwave listening. The elements of the record are usually date, time, frequency, callsign(s) plus any notations the SWL might want to make which would be helpful for future reference.

marker Repetition of a message or callup in Morse code or voice which identifies a station or serves to hold the frequency for future use.

MCW Modulated continuous wave, a radio signal in which the carrier is modulated by a constant audio-frequency tone. In telegraphy service, the carrier is keyed to produce the modulation.

message Any thought or idea expressed in plain or secret language, prepared in a form suitable for transmission by any means of communication.

message indicator A group of letters or numbers placed within an encrypted message to designate the keying elements applicable to that message.

mixed cipher alphabet A cipher alphabet in which the sequence of letters or characters in one or both of the components is not the normal sequence.

monoalphabetic substitution A type of substitution employing a single cipher alphabet by means of which each cipher equivalent, composed of one or more elements, invariably represents one particular plain text unit, wherever it occurs throughout any given message.

monograph A single letter.

monome A single digit.

monome-dinome system A substitution system in which certain plain text elements have single-digit cipher equivalents, while others are represented by pairs of digits.

Morse code A code developed by Samuel Morse which uses patterns of long and short pulses of energy to represent letters of the alphabet plus other characters.

multiliteral Of or pertaining only to crypto systems, cipher alphabets, and frequency distributions which involve cipher units of two or more letters or characters. See also *polygraphic.*

multipath Propagation of radio waves by more than one path from transmitter to receiver. Usually causes fading or distortion.

net An organization of radio stations which generally reflects the command structures. In a particular grouping of stations the one serving the senior echelon is usually the station in charge of the subordinate stations; this station is called the net control station, and the others are called outstations. The control station is responsible for the supervision of transmissions, procedures, and circuit discipline.

noise Any unwanted electrical disturbance or spurious signal which modifies the transmitting, indicating, or recording of desired data.

noncrashing A term used to describe that feature of the structure of certain cryptosystems which does not permit a plain text unit to be represented by the cipher text itself.

null In cryptography, a symbol or unit of encrypted text having no plain text significance.

numerical key A key composed of a sequence of numbers.

numerically key columnar transposition A columnar transposition system in which the columns of a matrix are taken off in the order determined by a numerical key.

off the cut A applied to the division of cipher text into polygraphs, beginning elsewhere than with the initial character of a bona fide polygraph.

one-part code A code in which the plain text elements are arranged in alphabetical, numerical, or other systematic order accompanied by their code groups also arranged in alphabetical, numerical, or other systematic order.

one-time pad A pad with a one-time use key printed on each page designed to permit the destruction of each page as soon as it is used. An agent in the field may have a pad the size of a postage stamp. Sometimes the pads are called "one-time gammas" or just "gammas."

one-time system A system of encipherment in which a non-repeating key is used.

one-time tape A tape used in the keying element in a one-time cryptosystem.

on the cut As applied to the division of text into polygraphs, beginning with the first textual character.

open code A cryptographic system using external text, which has an external meaning but disguised hidden meaning.

order of battle Abbreviated as OB. Basic information on ground-forces unit formation which provides all pertinent order-of-battle information.

originator The command by whose authority a message is sent. The originator is responsible for the functions of the drafter and releasing officer.

other end Usually refers to a station being called. Sample log entries would be: other end not heard; other end very weak; other end found 15 kHz below calling station frequency, etc.

out station The subordinate station(s) in a net.

overt Open, without attempt to conceal.

padding Extraneous text added to a message for the purpose of concealing its length and beginning or ending or both.

paraphrase To change the phraseology of a message without changing its meaning.

periodic polyalphabetic substitution A method of encipherment involving the cyclic use of two or more alphabets. Also called *repeating-key method.*

period system A system in which the enciphering process is repetitive in character and which usually results in the production of cyclic phenomena in the cryptographic text.

permutation table A table designed for the systematic construction of code groups. It may also be used to correct garbles in groups of code text.

phi test A test applied to a frequency distribution to determine whether it is monoalphabetic or not. See also *kappa plain* and *kappa random*.

physical security That component of security which results from all physical measures necessary to safeguard classified equipment and material from access by unauthorized persons.

piccolo A method of RTTY made up of tones. Initially introduced by the British Diplomatic Wireless Service. There are several forms of Piccolo with different number of tones in use by various countries.

pirates Unlicensed broadcasting stations run by electronic hobbyists.

plain component The sequence of plain text symbols in a cipher alphabet.

plaindress A type of message in which the originator and addressee designations are indicated outside the text and such designators are not enciphered.

plaintext Often abbreviated as PT. Of or pertaining to that which conveys an intelligible meaning in the language in which it is written with no hidden meaning, as the plaintext equivalents. Often shortened to *plain*.

polyalphabeticity A characteristic of encrypted text which indicates that it has been produced by more than one cipher alphabet. It is normally disclosed by frequency distributions which display "smoothness," or lack of pronounced variation in relative frequencies.

polyalphabetic substitution A type of substitution in which the successive plaintext elements of a message, usually single letters, are enciphered by a succession of different alphabets.

polygraphic substitution Encipherment by substitution methods in which the plaintext units are regular-length grouping of more than one element.

practice traffic As the term implies, traffic utilized in practice for transmitting, receiving and encryption/decryption functions.

probable-word method The method of solution involving the trial of plain text assumed to be present in a cryptogram.

propagation Manner in which an electromagnetic emission travels outward from its source.

Q-signals Used in communications to send questions and answers in an abbreviated form. Some common ones are:

QAP	Listen for me
QRA	The name of my station is ________
QRK	Signal readability
QRM	Interference (manmade)
QRN	Atmospheric noise (static)
QRU	I have nothing for you
QSA	Signal Strength
QSB	Your signals are fading
QSL	Acknowledgment
QSO	Two-way communication
QSW	I am going to transmit on (frequency)
QSX	I am listening to (frequency)
QSY	Shift to transmit on (frequency)
QSZ	Send each group twice
QTC	I have traffic for you
QTH	My location is ___________
QTR	The correct time is ___________

radio frequency Abbreviated rf, any frequency at which coherent electromagnetic radiation of energy is possible.

random 1. In mathematics, pertaining to chance variations from an expected norm. 2. In cryptanalysis, pertaining to any situation in which a statistical analysis will show variations from a calculated expected norm which variations are indistinguishable from those due to chance.

raw traffic Intercepted traffic showing no evidence or processing for communication intelligence purposes beyond sorting by clear address elements, elimination of unwanted messages, and the inclusion of an arbitrary traffic designator.

RCS Reduced carrier sideband. A modulation method in which only one sideband is transmitter but with enough remaining carrier so the signal can be listened to on ordinary radios not equipped to demodulate single sideband.

read To decrypt, especially as the result of successful cryptanalytic investigation.

relay A transmission forwarded by way of an intermediate action. Another station, other than the addressee, acts to pass the message from originator to addressee.

reversed standard cipher alphabet A cipher alphabet in which both the plain and cipher components are the normal sequence, the cipher component being reversed in direction from the plain component.

rotation systems Usually abbreviated as ROTA. Refers to a prescribed plan of systematic changing of callsigns and/or frequencies. One example of a callsign ROTA would be callsigns that changed daily and another would be callsigns that changed daily and were then repeated on the eighth day. This latter ROTA would be designated as "Daily changing/Weekly repeating." In a complex system involving both frequency and call ROTA plans, the callsigns used on different frequencies are also different. Thus one station might call ABC DE DEF on one frequency and be answered on another frequency by XYZ DE MNO. The purpose of such procedures is to create confusion for the Traffic Analysis efforts.

roughness A pronounced variation in relative frequencies of the elements considered in a frequency distribution.

route transposition A method of transposition in which the cipher-text equivalent of a message is obtained by transcribing, according to any prearranged route, the letters inscribed in the cells of a matrix into which the message was inscribed earlier according to some prearranged route.

scrambling See *ciphony.*

shortwave listening Abbreviated SWLing. The monitoring of transmissions (usually other than amateur) in the 1.8 to 30 MHz portion of the spectrum.

sigmage As used in cryptomathematics, a measure of the deviation from the normal, expressed in terms of numbers of sigmas.

signal intelligence (SIGINT) A generic term that includes both communications intelligence and electronic intelligence.

signal-to-noise ratio The ratio of the strength or a radio signal to that of noise.

simplex operation Operating method in which transmission is made possible alternately in each direction of a telecommunication channel, for example, by means of manual control.

SSB Single sideband. Radio transmission in which only the upper or lower band is transmitted and the carrier is completely suppressed.

single transposition A transposition in which only one inscription and one transcription are affected.

situation report (SITREP) The principal means of reporting to a higher authority information about the tactical situation and such administrative information as may affect the tactical situation.

sliding strip A strip of cardboard or similar material which bears a sequence and which can be slid against other such strips to various juxtapositions.

smoothness The lack of pronounced variation in relative frequencies of the elements considered in a frequency distribution.

solve To cryptanalyze. To find the plain text of encrypted communications by cryptanalytic processes, or to recover by analysis the keys and the principles of their application.

SPOT Report One-time reports used by all echelons to transmit intelligence or information of immediate value.

spread spectrum A signal structuring technique that employees direct sequence, frequency hopping or a hybrid of these, which can be used for multiple access and/or multiple functions. This technique decreases the potential interference to other receivers while achieving privacy and increasing the immunity of spread spectrum receivers to noise and interference. Spread spectrum generally makes use of a sequential noise-like signal structure to spread the normally narrow band information signal over a relatively wide band of frequencies. The receiver correlates the signal to retrieve the original information signal.

standard cipher alphabet A cipher alphabet in which the sequence of letters in the plain component is the normal, and in the cipher component is the same as the normal, but either reversed in direction or shifted from its point of coincidence with the plain component.

station call-up procedures 1. A frequent type of call-up is the double-station call procedure, wherein the call signs of the called station and of the transmitting station are sent, separated by DE. For example: ABC DE XYZ. The reply would be XYZ DE ABD. 2. In the single-station call procedure, only one call sign, usually that of the called station is used. 3. Sometimes one particular call sign is assigned to a link, i.e., for intercommunication between two specific stations. 4. A collective call sign is used for alerting all of the stations in the net. It is called a net call sign. 5. In all of the foregoing procedures, split-call working might be employed. A station when communicating with its superior uses one call sign, but when communicating with its own out-stations, it uses yet another call sign. These multiple call signs are used for convenience of operations, or for security. Another form is variant call signs with these usually being in a list from which the radio operator selects those to be used at any particular time.

stereotype A word, number, phrase, abbreviation, etc., which as a result of language habits, has a high probability of occurrence especially at the beginning or ending of a message.

strip-cipher system A polyalphabetic substitution system employing sliding strips in conjunction with a variable-generatrix principle.

stutter group Group consisting of a repeated single character, as 55555 or DDDDD, usually a four or five character group.

substitution system A system in which the elements of the plain or code text are replaced by other elements.

subtractive system A system in which encipherment is accomplished by subtracting the key from the plain text; in the process of decipherment, the enciphered text is added to the key.

superencipherment Result of subjecting cipher or coded text to a further process of encipherment.

syllabary In a code book, a list of individual letters, combination of letters, or syllables, accompanied by their equivalent code groups, usually provided for spell-out words or proper names not present in the vocabulary of a code; a spelling table.

syllabary square A cipher matrix containing individual letters, digits, syllables, frequent digraphs, trigraphs, etc., which are encrypted by the row and column coordinates of the matrix.

system indicator See *discriminant.*

telecommunication Any transmission, emission or reception of signs, signals, writing, images and sounds or intelligence of any nature by wire, radio, optical or other electromagnetic system.

telegraphy Usually refers to a form of telecommunication for the transmission of written matter by the use of a signal code.

telephony A form of telecommunication set up for speech (or other sounds) transmission.

teleprinter An electrically-operated instrument used in the transmission and reception-printing of messages by proper sensing and interpretation of electrical signals.

text The part of a message containing the basic information which the originator desires to be communicated.

traffic Communications volume during a given period. Abbreviated as *tfc*.

traffic analysis That branch of cryptology which deals with the study of the external characteristics of signal communications and related materials for the purpose of obtaining information concerning the organization and operation of a communication system.

transcription 1. In a transposition system, the process of removing the text from a matrix or grid by a method or route different from that used in the inscription. 2. A written copy of a previously recorded radio transmission; also the process of preparing such copy from tapes or records.

transmission security That component of communication security which results from all measures designed to protect transmissions from interception, traffic analysis, and imitative deception.

transmitter Equipment used to generate and amplify an rf carrier signal, modulate this carrier with intelligence, and radiate the modulated rf carrier into space.

transposition system A cryptosystem in which the elements of plain text, whether individual letters, groups of letters, syllables, words, phrases, sentences, or code groups or their components undergo some change in their relative positions without a change in their identities.

trigraph A set of three letters.

triliteral frequency distribution A distribution of the characters in the text of a message in sets of three, which will show: (a) each character with its two preceding characters; or (b) each character with its two succeeding characters; or in its most usual form, (c) each character with one preceding and one succeeding character. A triliteral frequency distribution of ABCDEF would consider the groups ABD, BCD, CDE, DEF.

trinome A set of three digits.

tripartite alphabet A multiliteral alphabet in which the cipher units may be divided into three separate parts whose functions are clearly defined, viz., page, row, and column indicators of a dictionary system.

triplet A group of three like symbols.

trough In its cryptologic application, a point of low relative frequency in a frequency distribution.

two-part code A randomized code, consisting of an encoding section in which the plaintext groups are arranged in an alphabetical or other systematic order accompanied by their code groups arranged in a nonalphabetical or random order; and a decoding section, in which the code groups are arranged in alphabetical or numerical order and are accompanied by their meanings as given in the encoding section.

utility communications The utility classification is generally considered to include all communications except amateur and broadcast transmissions.

variant system A substitution system in which some or all plaintext letters may be represented by more than one cipher equivalent.

Vigenere square The cipher square, commonly attributed in cryptographic literature to the French crypto expert, Blaise de Vigenere. It consists of a square having the normal sequence forming the successive rows or columns within the square. The term is sometimes applied to a square exhibiting such symmetry but with a mixed sequence.

white noise Broadband noise of equal amplitude at all points.

word pattern The characteristic arrangement of repeated letters in a word which tends to make it readily identifiable when enciphered monoalphabetically.

word separator A unit of one or more characters employed in certain cryptosystems to indicate the space between words. It may be enciphered or unenciphered. Also called a *word spacer.*

word transposition A cryptosystem in which whole words are transposed according to a certain prearranged route or pattern.

working Stations communicating with each other. In simplex working, stations operate on a common frequency; in complex working, more than one frequency is used.

About the Author

Don Schimmel has a broad background in communications covering almost fifty years, including 30 years of combined military and government service.

During a four year Navy hitch, Don served in the Philippines, China, Korea, and various south Pacific islands. After his discharge from the Navy in 1948, he attended the Milwaukee School of Engineering. From 1951 to 1977, he was employed by the U.S. federal government and served in a variety of communications-related positions, some of which involved liaison with various government components. During his federal career, he spent ten years in Central and South America and also went on official duty travel to Africa, the Far East, the Middle East, and Latin America.

Don writes the monthly utility column "Communications Confidential" in ***Popular Communications*** magazine and has had articles published in the annual Popular Communications ***Communications Guide***. He has also written columns for ***Monitoring Times, Umbra et Lux***, and the ***Numbers Factsheet***. From time to time, he has contributed articles to the Association of DX Reporters club newsletter and to the Universal Radio ***RTTY Listener***. For a change of pace, he writes an occasional genealogical article for a family association newsletter.

Don's initial shortwave interest was listening to shortwave broadcasts, but as time went on he became more and more interested in monitoring utility transmissions, particularly the mysterious signals found throughout the spectrum.

Being retired, Don generally has ample time for monitoring. His present equipment consists of a Drake R-8 receiver, two Kenwood R-2000 receivers, Grove Minituners, a homemade antenna switching panel, a frequency counter, video monitor, Universal M-7000 demodulator, Realistic PRO-43 scanner, cassette tape recorder, and a Seikosha SL-80AE printer. For his antenna farm, Don has a 30-foot vertical whip and three towers which support various HF dipoles and longwires plus a VHF/UHF scanner antenna.